Issued under the authority of the Home Of̶

# Manual

# Firemans

A survey of the science of firefight. ̶

# Book 5
# Ladders and
# Appliances

London
Her Majesty's Stationery Office

First published 1984

The structure and publishing history of
the *Manual* is shown on pages 145–8

ISBN 0 11 340585 5

# Preface

One of the principal requirements of the fireman is equipment to help him effect rescues when he cannot do so unaided, gain access to buildings and other places, and occupy the best position for fighting a fire. The equipment he uses for this purpose is the ladder. There are various types that have been developed for Fire Brigade use, both portable and permanently mounted, hand-operated and mechanically or hydraulically operated.

This Book of the *Manual* looks first at the four main classes of ladder. Part 1 deals with three kinds of portable, hand-operated, ladders in current use: the extension ladder in its different sizes, the roof ladder, and the hook ladder (the last of these is now used by one large Brigade only). Part 2 looks at wheeled escapes. Formerly one of the main Fire Brigade ladders for rescue purposes, the escape, has now been largely superseded by the extension ladder, but a number are still in use in various Brigades. Part 3 looks at turntable ladders, the permanently mounted ladders designed to reach greater heights than either escapes or extension ladders and to be usable as water towers. Part 4 of the Book looks at hydraulic platforms. Strictly speaking, of course, these are not ladders, but they are dealt with here since they share a number of the same characteristics and are used for the same sort of purpose.

These various pieces of equipment are complementary. Each has its advantages and disadvantages and none can wholly replace another as best in all circumstances. Within financial and other constraints Brigades therefore choose their equipment according to their own preferences and taking into account any direct practical experience and the local conditions. These Parts of Book 5 are intended as a general guide. They describe the main characteristics common to the different pieces of equipment in each class, look more closely at some of the types at present in common use, and discuss the general principles of working with them.

In the Fire Service appliances are usually classed as 'pumping' appliances or as 'special' appliances. The latter expression applies to all vehicles except normal pumping appliances and covers a very wide range. None of these specials has a J.C.D.D. specification except for the emergency tender. Part 5 looks at some of the more common types in use by Brigades.

Part 6 deals with the 'work-horse' of the Service, the pumping appliance, and discusses some of the points to bear in mind when working with them. The practical operation of the pumps mounted on these vehicles is dealt with in the *Manual*, Part 2.

References in this Book to the male person should be construed as applying, where appropriate, to the female person also. The qualifications for appointment to and promotion within the Fire Service set out in the *Fire Service (Appointments and Promotion) Regulations 1965* applied only to men. By virtue of the *Fire Service (Appointments and Promotion) (Amendent) Regulations 1976* they now apply equally to women. References to firemen and other male holders of Fire Service rank should therefore be interpreted as references to firewomen and female holders of such rank also.

The Home Office is indebted to all who have helped in the preparation of this work.

Home Office
1984

# Metrication

List of SI units for use in the Fire Service

| Quantity and basic or derived SI unit and symbol | Approved unit of measurement | Conversion factor |
|---|---|---|
| **Length**<br>metre (m) | kilometre (km)<br>metre (m)<br>millimetre (mm) | 1km = 0.621 mile<br>1m  = 1.093 yards<br>     = 3.279 feet<br>1mm = 0.039 inch |
| **Area**<br>square metre (m²) | square kilometre (km²)<br>square metre (m²)<br>square millimetre (mm²) | 1km² = 0.386 mile²<br>1m²  = 1.196 yards²<br>      = 10.764 feet²<br>1mm² = 0.002 inch² |
| **Volume**<br>cubic metre (m³) | cubic metre (m³)<br>litre (1) (= 10⁻³m³) | 1m³ = 35.7 feet³<br>1 litre = 0.22 gallon |
| **Volume, flow**<br>cubic metre per second (m³/s) | cubic metre per second (m³/s)<br>litres per minute (1/min) | 1m³/s = 35.7 feet³/s<br>1l/min = 0.22 gall/min |
| **Mass**<br>kilogram (kg) | kilogram (kg)<br>tonne (t) | 1kg = 2.205 lbs<br>1t = 0.984 ton |
| **Velocity**<br>metre per second (m/s) | metre per second (m/s)<br>international knot (kn) (= 1.852km/h)<br>kilometre per hour (km/h) | 1m/s = 3.281 feet/second<br>1km/h = 0.621 mile/hour |

| Quantity and basic or derived SI unit and symbol | Approved unit of measurement | Conversion factor |
|---|---|---|
| **Acceleration**<br>metre per second²(m/s²) | metre/second² (m/s²) | $1m/s^2 = 3.281$ feet/second² $= 0.102$ 'g' |
| **Force**<br>newton (N) | kilonewton (kN)<br>newton (n) | $1kN = 0.1$ ton force<br>$1N = 0.225$ lb force |
| **Energy, work**<br>joule (J)<br>(= 1Nm) | joule (J)<br>Kilojoule (kJ)<br>Kilowatt/hour (kW/h) | $1kJ = 0.953$ British Thermal Unit<br>$1J = 0.738$ foot lb force |
| **Power**<br>watt (W)<br>(= 1J/s = 1Nm/s) | kilowatt (kW)<br>watt (W) | $1kW = 1.34$ horsepower<br>$1W = 0.735$ foot lb force/second |
| **Pressure**<br>newton/metre² (N/m²) | bar (= $10^5$N/m²)<br>millibar (mbar)<br>(= $10^2$N/m²)<br>metrehead<br>(= 0.0981 bar) | 1 bar = 0.991 atmosphere<br>$= 14.5$ lb force/in²<br>1mbar = 0.0288 inch Hg<br>1 metrehead = 3.28 foot head |
| **Heat, quantity of heat**<br>joule (J) | joule (J)<br>kilojoule (kJ) | $1kJ = 0.953$ British Thermal Unit |
| **Heat flow rate**<br>watt | watt (W)<br>kilowatt (kW) | $1W = 3.41$ British Thermal Units/hour<br>$1kW = 0.953$ British Thermal Unit/second |
| **Specific energy, calorific value, specific latent heat**<br>joule/kilogram (J/kg)<br><br>joule/m³ (J/m³) | kilojoule/kilogram (kJ/kg)<br>kilojoule/m³ (kJ/m³)<br>megajoule/m³ (MJ/m³) | $1kJ/kg = 0.43$ British Thermal Unit/lb<br>$1kJ/m^3 = 0.0268$ British Thermal Unit/ft³ |
| **Temperature**<br>degree Celsius (°C) | degree Celsius (°C) | 1 degree Celsius = 1 degree Centigrade |

# Contents

## Part 1
## Extension, roof and hook ladders

# Part 2
# Escapes

# Part 3
# Turntable ladders

# Part 4
# Hydraulic platforms

## Chapter 13 Working with hydraulic platforms

# Part 5
# Special appliances

1. Introduction

## Chapter 14   Special appliances

1 Emergency tenders
   a. Specifications
   b. Types of emergency tenders
   c. Equipment
   d. Examples

2 Water carriers

3 Breakdown lorries

4 Decontamination/Chemical incident units

5 Foam tenders

6 Rolonoff units, pods, caravans etc.
   a. Types
   b. Operational considerations

7 Control units

8 Hose laying lorries

# Part 6
# Pumping appliances

# List of Plates

1   Metal three-piece short extension ladder.
    *Photo: London Fire Brigade*

2   10.5m metal extension ladder.
    *Photo: London Fire Brigade*

3   10.5m wooden extension ladder.
    *Photo: Hampshire Fire Brigade*

4   13.5m metal extension ladder.
    *Photo: London Fire Brigade*

5   Roof ladder.
    *Photo: London Fire Brigade*

6   A 13.5m ladder being used to rescue a man trapped in
    defective window-cleaning cage.
    *Photo: Hampshire Fire Brigade*

7   A typical detachable two-man cage fitted at the head of a
    turntable ladder.
    *Photo: Avon Fire Brigade*

8   An illuminated Field of Operations Indicator attached to the
    main ladder of a turntable ladder.
    *Photo: Grampian Fire Brigade*

9   Magirus DL30 turntable ladder with cage carried on
    turntable.
    *Photo: Oxfordshire Fire Service*

10  Main controls of Magirus DL30 turntable ladder. Note the
    field of operations indicator top left.
    *Photo: Avon Fire Brigade*

22 Metz DL30 turntable ladder. By use of blocks under the jacks the angle of depression has been increased to –25° maximum
*Photo: Grampian Fire Brigade*

23 Pitching a turntable ladder across the chassis. In areas like this, care must be taken that jacks are not placed on manhole covers, drain-gratings etc.
*Photo: North Yorkshire Fire Brigade*

24 A Simon SS 220 hydraulic platform with jacks fully extended. Note the position of the cage in the housed position.
*Photo: London Fire Brigade*

25 A Simon hydraulic platform illustrating the position of the cage fully housed. Note the fixed piping running along the upper boom and the water collecting head at the rear off-side.
*Photo: Isle of Wight Fire Brigade*

26 Simon SS 220 hydraulic platform cage fitted with folding platform monitor, hose connections, lighting etc.
*Photo: London Fire Brigade*

27 Simon SS 220 hydraulic platform. Control console and boom controls.
*Photo: London Fire Brigade*

28 Simon SS 220 hydraulic platform showing cage controls and the general construction of the cage.
*Photo: London Fire Brigade*

29 At low elevations the knuckle can project a considerable distance. Hydraulic platform operators must remember this when training the apparatus.
*Photo: South Glamorgan County Fire Service*

30 The cage operator's ability to manoeuvre the cage over a building at the limit of the boom is a distinct advantage.
*Photo: London Fire Brigade*

31 An electrical control panel on a type "A" emergency tender.
*Photo: Avon Fire Brigade*

32  Air bags, inflated from a breathing apparatus cylinder, being used to lift a vehicle.
*Photo: Bedfordshire Fire Service*

33  A typical metropolitan type "A" emergency tender. The electrical control panel is located on the rear side of the appliance behind the shutters.
*Photo: London Fire Brigade*

34  A type "B" emergency tender. This particular vehicle also acts as a BA Tender and a Control Unit. A portable generator can be seen in the rear off-side locker.
*Photo: Cornwall County Fire Brigade*

35  In areas where there is difficulty in obtaining water, water carriers, such as this example, are invaluable. Carrying 4,500 litres of water and a light portable pump it also has the ability to cross rough terrain.
*Photo: Cornwall County Fire Brigade*

36  A breakdown lorry with the facility of lifting and winching 7.5 tonne. Fitted with stabilising jacks, it also carries a range of jacks, air-bags, cutting equipment etc.
*Photo: Durham County Fire Brigade*

37  A combined decontamination/chemical incident unit which, following decontamination in an area set out at the back, allows personnel to shower, dress and exit at the front. Airline equipment can be seen at the rear near-side.
*Photo: West Yorkshire Fire Service*

38  The decontamination area of Plate 37. Warning signs, spare cylinders, dam, vacuum cleaner and airlines are visible with clear directions to personnel.
*Photo: West Yorkshire Fire Service*

39  A county foam tender showing the range of gear stowed. 900L of protein foam compound and 1800L of water are also carried and the appliance is equipped with two pumps plus HEF generators.
*Photo: County of Clwyd Fire Service*

40 A metropolitan county foam tender with a fixed monitor. In addition to 3000L of foam compound, 450L of HEF compound and a Turbex generator are carried.
*Photo: London Fire Brigade*

41 A Rolonoff unit showing the pod ready for loading onto the prime mover. This particular pod is a decontamination unit which can be set down and deployed where necessary leaving the prime mover free to be used to move another pod.
*Photo: South Yorkshire County Fire Service*

42 A demountable pod with electro-hydraulically operated legs which can be extended before the prime mover is removed. This particular pod is a control unit.
*Photo: North Yorkshire Fire Brigade*

43 An adaption of a 4-wheeled caravan to a decontamination unit to be towed into position as necessary.
*Photo: Cleveland County Fire Brigade*

44 A divisional control unit used at smaller incidents.
*Photo: London Fire Brigade*

45 A comprehensively equipped metropolitan fire brigade control unit. The masts are either retractable or folding and the interior is sub-divided into three compartments (see Plate 46).
*Photo: London Fire Brigade*

46 The interior of the appliance shown in Plate 45, taken from the communications compartment. Beyond is the working area and, at the far end, the conference section.
*Photo: London Fire Brigade*

47 A shire county brigade control unit. This is divided into a radio compartment and a working area (see Plate 48)
*Photo: Dorset Fire Brigade*

48 The interior of the appliance shown in Plate 47. Taken from the radio compartment and showing the county map, mobilising board and perspex working surfaces.
*Photo: Dorset Fire Brigade*

49   A typical hose-laying lorry carrying 1,850m of 90mm hose
     and able to lay single or double lines at speeds of up to
     50 kph.
     *Photo: West Yorkshire Fire Service*

50   Example of pump side controls to a rear mounted pump,
     showing gauges, hose-reel valves, high and low pressure
     controls, tank filling facility and the usual inlets and outlets.
     *Photo: London Fire Brigade*

51   An L4P equipped with a 2250 l/min front-mounted pump.
     This type of appliance is very suitable for an area with rough
     terrain and narrow roads.
     *Photo: Cornwall Fire Brigade*

52   A Chelsea side power take off fitted as a "sandwich" type.
     *Photo: Shelvoke and Drewry Ltd*

53   A power take off, "drive-line" type, fitted between the gearbox
     and axle,
     *Photo: HCB-Angus Ltd*

54   The chassis of an appliance showing the arrangement of
     water tank, power take off and rear-mounted pump with side
     controls.
     *Photo: Hestair Dennis Ltd*

55   Brigades vary in their appliance stowage arrangements.
     Considerable ingenuity is shown in carrying a considerable
     amount of equipment, readily available, in a comparatively
     small space.
     *Photo: West Yorkshire Fire Service*

56   Another example of stowage design. Firemen should know the
     stowage on their appliance but clear labelling obviously
     helps.
     *Photo: Hertfordshire Fire Brigade*

57   A dual-purpose appliance carrying a 9m ladder, 1365 l of
     water and a rear-mounted pump with side controls. The
     trunnion bar at the back and quickly removable gantries
     enables rapid conversion to a pump-escape.
     *Photo: London Fire Brigade*

58  A water-tender ladder carrying 13.5m, 9m, short extension
    and roof ladders, 1,800l of water and a 4,500l/min rear-
    mounted pump. Based on a Dennis R133 chassis and driven
    by a V8–640 162KW diesel engine.
    *Photo: Derbyshire Fire Service*

59  Another example of a water-tender ladder on an HCB–Angus
    chassis. Carrying 1800l of water, a 13.5m ladder and a
    2,250l/min pump, this appliance is also equipped with a
    Stemlite telescopic lighting mast.
    *Photo: Staffordshire Fire Brigade*

60  A "compact" appliance based on a Ford "A" series chassis.
    The tank holds 675l of water. Pumping capacity 2270l/min
    multi-pressure plus 2 hose-reels, 10.5m, short extension and
    roof ladders, 3 BA sets and a crew of five men.
    *Photo: Cumbria Fire Service*

61  A Simonitor appliance capable of delivering foam or water at
    an extension of 12.6m on a turntable. It carries 1,800l of
    foam concentrate, 135l of HEF and an HEF unit. There are
    two jacks at the rear to stabilise the appliance as it operates.
    *Photo: Essex Fire Brigade*

62  An experiment to bring the weight of a water-tender ladder
    below 7.5t. Based on a Bedford KD 120 chassis and powered
    by a 4900cc diesel, it carries a Godiva 50 multi-pressure
    pump and 1125l of water. Aluminium is extensively used in
    the construction.
    *Photo: Strathclyde Fire Brigade*

63  A pump-escape carrying a 15m steel escape on a Miles
    mounting. Water tank capacity 1350l and fitted with a
    Godiva UMP pump.
    *Photo: East Sussex Fire Brigade*

# Part 1
# Extension, roof and hook ladders

## Introduction

Ladders which are carried into position by hand are an important part of firemen's equipment. It is essential that they understand their construction and use. Apart from roof and hook ladders they all consist of a main ladder plus one or two extensions. The extensions are raised either by hauling on a line or by pushing up by hand. The ladders are carried on different appliances in various combinations according to local needs.

The main advantage of all these ladders is that firemen can use them in awkward positions, e.g. along narrow alleys, over walls, or under arches at the back of premises. These would be inaccessible to the wheeled escapes, turntable ladders and hydraulic platforms described in later Parts of this Book. There has in fact been for some time an increasing tendency to replace escapes completely with extension ladders. Firemen can also use the smaller ladders in numerous other ways, e.g. bridging both inside and outside buildings, replacing destroyed staircases, descending into basements, or as stretchers. They can separate the sections of an extension ladder for single use or can use it to improvise a step-ladder.

This Part of Book 5 looks at the general characteristics of the different ladders that are carried into position by hand and considers the broad principles of working with them.

# Chapter 1
# General information on extension, roof and hook ladders

## 1  Terminology

Firemen should first ensure they are familiar with the proper terms for the various parts of a ladder as indicated in Fig. 1.1. They should note in particular that the Fire Service always refers to the rungs of a ladder as the rounds. Firemen should also familiarise themselves with the following terms used in operations with these ladders:

extend — to raise the extending portion of a ladder;

extend to lower — to raise the extending portion of a ladder to clear the pawls for lowering;

heel in, out — to move the heel of a ladder towards or away from a building;

lower — to retract the extending portion of a ladder;

pitch — to erect a ladder against a building;

slip — to remove a ladder from an appliance.

## 2  Specifications

The Joint Committee on Design and Development of Appliances and Equipment of the Central Fire Brigades Advisory Councils for England and Wales and for Scotland has laid down specifications for various sizes of extension ladder, roof ladders and hook ladders. Firemen can expect most ladders in local authority Fire Brigades to comply with these. The following specifications currently apply:

JCDD 32   Short extension ladders;

JCDD 37   nine metre extension ladders;

JCDD 10   10.5 metre extension ladders;

JCDD 12   13.5 metre extension ladders;

JCDD 31   roof ladders

JCDD 11   hook ladders

The specifications allow ladders to be of wood or of metal, provided they can pass any acceptance test laid down in the specifications.

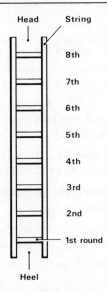

Head    String

8th

7th

6th

5th

4th

3rd

2nd

1st round

Heel

Fig. 1. The principal parts of a ladder.

Wooden ladders may be of solid or laminated construction. They should be of straight-grained, well-seasoned timber, free from defects and should be clear varnished so that defects are not hidden. Where a length is specified for the ladder, this refers to the ladder's length when fully extended. This Chapter gives only a general description of ladders complying with the various specifications. Any fireman who wants more detailed knowledge must consult the specifications themselves.

## 3    Extension ladders

### a. Short extension ladders

These consist of two or three extensions, most modern ladders being made of metal (see Plate 1). JCDD 32 is based on BS 2037, *Aluminium Ladders* but incorporates additional requirements to render the ladders suitable for Fire Brigade use. Fully extended, these ladders should be between 5.5*m. and 6.7 m. long. They should not weigh more than 24.6 kg and the space from round to round should be between 280 mm and 305 mm. They are design to meet various criteria, including degree of sway when pitched, overlap, deflection and speed of separation for use as individual sections. The last is particularly important since it is probably this ladder which firemen use for the widest variety of purposes (see Introduction).

*NB All measurements in this Book are rounded to a convenient figure unless stated to be exact.

## b. Nine metre ladders

Firemen have used ladders of about this length for a long time, though JCDD 37 dates only from 1978. The 10.5 m ladder had been generally superseding them but there appears to have been a recent tendency back to the smaller size. The following are among general features of the ladders:

(i)     reasonable rigidity and freedom from sway under prescribed conditions of pitch and load;

(ii)    one main and one extending section, the latter having a width inside the strings of not less than 300 mm;

(iii)   easy extension by one man by an endless rope over a pulley or other approved means;

(iv)   safety devices, including pawls of an approved design and non-skid shoes on the heels of the main ladder;

(v)    a weight not exceeding 56kg;

(vi)   operating gear so designed that it is easy to remove the extension;

(vii)  the ability to pass stringent deflection and round tests.

Within the requirements of the specification the actual construction of the ladder varies from manufacturer to manufacturer. Figs 1.2 and 1.3 show examples, respectively, of a typical metal ladder and a typical wooden ladder.

### (1) Metal ladders

Typically these are of a riveted trussed construction with high tensile aluminium alloy extrusions and square section rounds ribbed on the tread surface to make them less slippy. The extending section runs on nylon rollers in guide channels integral with the stiles and double action automatic pawls are fitted. The line is 16 mm in diameter and rot proofed. The head of the ladder usually has wheels for easy running on wall surfaces. Some ladders are made without riveting and have rollers on the head instead. Pawls, heel protection, pulleys and lines vary.

### (2) Wooden ladders

Most wooden ladders are trussed. The trussing has to be on the underside of the ladder when it is in operational use. Pawls seldom operate automatically, except when the ladder is being housed; they are usually cleared initially by extending to lower.

## c. 10.5m ladders

JCDD 10 (issued in 1974 and amended 1979) lays down essentially the same requirements as for the nine metre ladder (see Section 3b

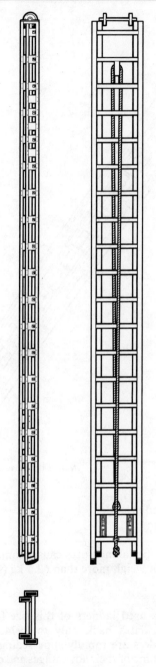

Fig. 1.2  A 10.5 m metal ladder.

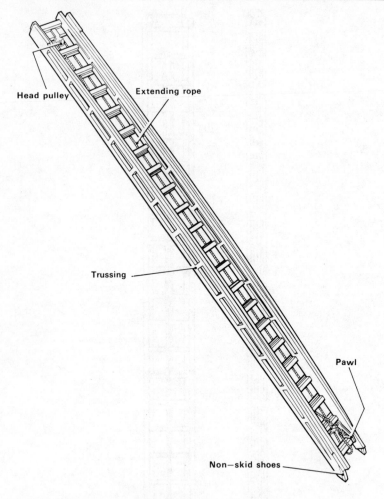

Fig. 1.3 A 10.5 m wooden ladder.

above). The use of Imperial measures causes some slight variations. The ladder should not weigh more than 63.5 kg (see Plates 2 and 3).

### d. 13.5m ladders

The Fire Service has used ladders of this size for about 25 years though JCDD 12/1 dates back only to 1974. As noted in the Introduction the ladders are rapidly superseding wheeled escapes. Chapter 2 discusses some of their advantages and disadvantages, and Fig 1.4 shows a typical example.

The specification differs in several areas from those for nine metre and 10.5 m ladders. The main difference is that the 13.5 m ladder has

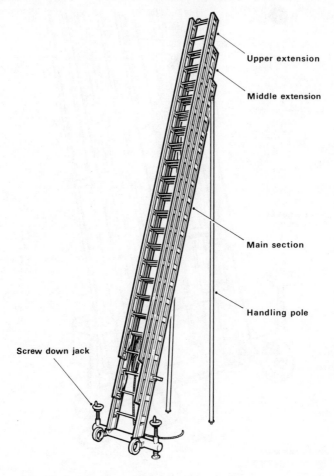

Upper extension

Middle extension

Main section

Handling pole

Screw down jack

Fig. 1.4 A 13.5 m ladder

one main section and two extensions, the width of the narrowest being not less than 305 mm. The ladder also has suitable props with universal joints. These props are attached to the main ladder and have non-skid attachments on their feet. They must stow snugly alongside the strings when not in use (see Plate 4).

The ladder must also have quadrants or spikes to help in elevation and non-slip adjustable jacks for plumbing (see Fig. 1.5). Their position must be such that they allow room for a fireman's boot between the jack and the main ladder string. The extra extension is operated by a cable which must meet various requirements for approved fixing, minimum breaking load (1.3 tonnes), sheaves etc. Exact weight limits for the ladder are: maximum, 113 kg; preferable, 102 kg.

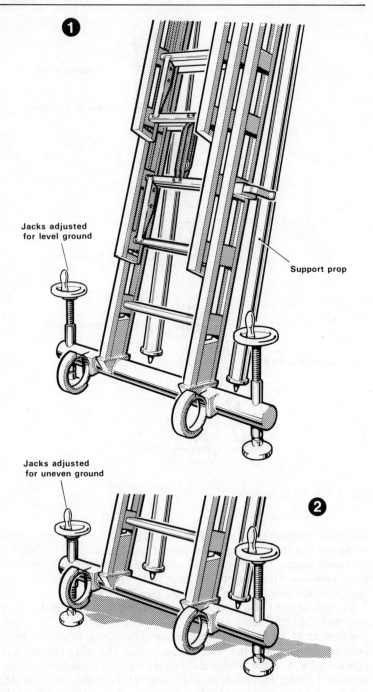

**1**

Jacks adjusted
for level ground

Support prop

Jacks adjusted
for uneven ground

**2**

Fig. 1.5 Adjustable jacks on a 13.5m ladder

## 4    Roof ladders

Roof ladders which comply with JCDD 31 are light and strong, can be readily handled by one man, and are suitable for securing by means of a hook which fits over the ridge of the roof. When in use the strings are supported along their length by roof-bearers. The ladders are usually of a one-piece construction (see Plate 5) and not less than 4.5 m long. A ladder longer than 5.4 m can be two piece or folding. Toe clearance is obviously necessary and is not less than 50 mm. Two other important requirements which the specification lays down are non-slip treads on the rounds and two wheels of not less than 100 mm diameter on the upper side of the strings or incorporated with the ridge hook. The ladders weigh no more than 16 kg for a one-piece and 18 kg for a folding or two piece.

## 5    Hook ladders

This ladder (see Fig. 1.6) works on an entirely different principle from all other Brigade ladders. It must be suitable for suspension by means

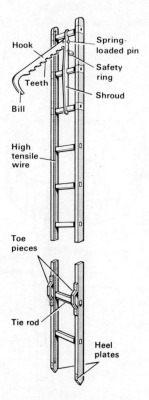

Fig. 1.6 Hook ladder showing details of parts

of a hook from a window sill so that one man can ascend or descend unassisted. Ladders complying with JCDD 11 weigh no more than 13.2 kg and have a total length of 4.1 m. Because of their special nature they have to meet very strict and detailed requirements, with regard, for example, to:

(i)     materials and their finish;

(ii)    type of hook, anchorage, and safety ring;

(iii)   protection and safety: where necessary, for example, metal shoes, tie rods under certain rounds, high tensile wire reinforcement;

(iv)    toe pieces to provide a 75 mm clearance from the wall;

(v)     balance

The requirements vary between ladders made of wood and those made of metal.

### a. Hook ladder belts

Firemen should wear a hook ladder belt whenever they use a hook ladder. The belt in general use is illustrated in Fig. 1.7. It is about 115 mm wide and just over one metre long, made of best quality webbing lined with leather. It is fastened with two straps which buckle from

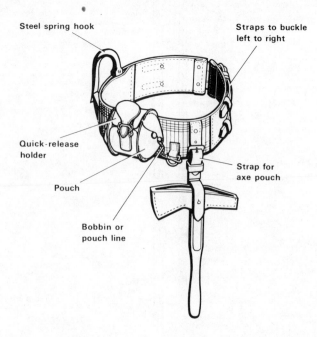

Fig. 1.7  Hook ladder belt.

left to right to obviate loose ends of the straps interfering with the spring hook. A large spring hook is fixed to the belt and is used to hook onto the safety ring fitted to the head of the hook ladder. The belt also has a bobbin line pouch which contains 40 m of plaited cord (see *Manual,* Book 2, Chapter 11) and possibly an axe pouch.

## 6   Acceptance tests

The JCDD specifications lay down acceptance tests for all sizes of extension ladder. These consist of a deflection test and a round test, plus a side test for the 13.5 m ladder. The tests are carried out at the works, the Brigade or elsewhere.

The other ladders are also subject to acceptance tests. Those for the hook ladder, especially on its hook, anchorage and safety ring, are very stringent. In some cases they require marked British Standard materials accompanied by a manufacturer's test certificate. Other acceptance tests for hook ladders comprise ones on ladder deflection, rounds and balance, and a general test.

# Chapter 2
# Working with extension, roof and hook ladders

## 1 Handling ladders

The ability to handle ladders safely and quickly under adverse conditions is the hallmark of a good fireman. The contrast between slipping, manoeuvring and pitching a ladder on a calm, clear day with plenty of room and on firm ground and the same operations attempted about midnight, in rain, a strong wind, up a narrow alley, amongst industrial debris, is obvious. Firemen should practise unusual pitches under whatever unusual conditions they can simulate. It is especially important that they practise working in darkness or at least with the illumination of appliance searchlights only, as this would often be the case at incidents.

Carrying ladders safely, especially under adverse conditions, requires skill. The *Fire Service Drill Book* describes the basic methods, but firemen should remember that, at large fires, they may have to step over numerous lines of hose, pass ladders across walls, negotiate narrow openings and twisting passages, and so on. They should even practise carrying a ladder down a slope. They must discover the problems they are likely to face and evolve satisfactory methods of dealing with them. It is especially important that the four men carrying a 13.5 m ladder should balance its weight equally amongst themselves. Any sudden transfer of weight could cause injury.

## 2 Hoisting a ladder

Firemen may find it necessary to haul up a ladder or section of a ladder to an upper floor or flat roof. The easiest and safest method is to haul up the ladder vertically by line, preferably with a guy line attached to keep it clear of projections. If the hoisting line is tied to the ladder approximately one third of its length from the top, this can help when the ladder is levered over a sill, balcony rail or roof edge. Fig. 2.1 (1) illustrates a method of attaching the line one third of the way down, Fig. 2.1 (2) at the top. Another method, where the distance is not too great, is to use the centre of a line for the knot. This allows the lower part to act as a guy line.

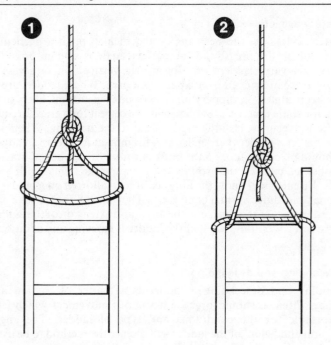

Fig. 2.1 (1) Hoisting a ladder with a line by tying a bowline one third down.

Fig. 2.1 (2) Hoisting a ladder with a line by a bowline tied at the top.

## 3   Pitching, climbing and working on an extension ladder

### a. General

The notes on ladder drills in the *Fire Service Drill Book* give general advice on the pitching of ladders. Firemen should study the *Drill Book* and follow its advice as far as they can. Where they pitch a ladder depends on circumstances. A relatively wide opening with a firm sill would enable firemen to place their ladder in one side of the opening with two or three rounds above the sill. This would leave plenty of room to get in and out, safe handholds on the ladder, space to maneouvre rescued people, room to suspend hose alongside the ladder, etc. With a narrow opening, firemen would have to pitch below the sill or to one side to leave enough room. Similarly, the same pitch may be necessary where, because of ground hazards, the ladder is at the extreme limit of its extension.

13

## b. Safety aspects

Firemen will know the best angles at which to pitch a particular ladder but on the fireground the perfect angle is seldom possible. Angles will vary: ladders may be nearly vertical where access is narrow, or semi-bridged over a basement area. Where ladders are at fairly steep angles, on slippery surfaces or uneven ground, likely to be in situ for some time, or providing a line of retreat for a crew, firemen should, for security, lash the ladder head to the building. Whenever anyone is on a ladder there must be a fireman at its heel to steady it. He should place one foot on the bottom round or jack beam, bracing the other leg back on the ground and grasping the strings with both hands. Firemen will know the number of men allowed on a ladder at any one time under normal conditions. They must however take into account factors at the time, e.g. the angle and extension of the ladder, the ground conditions, whether the ladder is carrying any additional weight such as hose.

## c. Ascending and descending

Ascending and descending a ladder is a matter of rhythm and balance. Firemen should practise a smooth movement which feels comfortable for the conditions and type of ladder. The angle, extension and 'whip' of the ladder will dictate the method to be used. Where the ground is firm and the ladder at its optimum angle, the fireman should raise his hands and feet in unison (i.e. left with left, right with right); he should keep his arms fairly straight, his body away from the ladder, his head slightly angled up to see where he is going and his feet sufficiently apart to give balance on that particular ladder (see Fig. 2.2). As a general guide the fireman should grasp the round level to his chest with palms down and thumbs underneath. Obviously if the ladder is angled steeply, leaning back to the proper angle may be impossible owing to lack of space or may tend to pull the ladder from the building. Conversely, at a shallow angle, the fireman must bend his knees and arms and be more careful not to make the ladder roll.

## d. Taking a leg-lock

Often when working on a ladder a fireman will need both hands free for such operations as directing a jet, handling small gear or pitching a hook ladder. In these circumstances he must first take a leg-lock. To do so he should grasp an appropriate round with both hands and place one leg through the ladder over the second round above that on which he is standing. He should bend this leg using the inside of his knee and back of his calf to 'lock' himself on. If possible he should also bring the foot of this leg back through the rounds, placing his toes under the round immediately above the one on which he is standing (Fig. 2.3). If working to the right he should use his left leg to make the lock, if to the left, his right.

Fig. 2.2 Correct method of climbing a ladder.

Ladder to
be footed

Fig. 2.3 Taking a leg-lock.

15

# 4   Using an extension ladder

## a. The 13.5 m ladder

This ladder has been used by the Fire Service for about 25 years and, as already noted, is now rapidly superseding the wheeled escape. Its obvious advantage is that it can be carried into places otherwise inaccessible and manoeuvred in confined spaces. With the use of props it becomes a steady platform for firefighting or rescue work (Plate 6). It does require a full crew of four for safe pitching and slipping, whereas in an emergency two can slip and pitch an escape. Present day conditions of car-parking, precincts etc., however favour a non-wheeled piece of apparatus.

## b. Other extension ladders

As already noted, extension ladders can be put to a variety of uses; the short extension ladder being particularly versatile.

### (1) Use as a stretcher

When a stretcher is not available firemen can use one section of a short extension ladder for carrying an injured person. They should ensure they are practised in this operaton and, for the following reasons, the practice should involve the use of a live body:

(i)     it will teach them to secure the body properly to the ladder: neither a dummy nor an unconscious person can explain if the lines are too tight or too loose or the ladder unpadded and uncomfortable;

(ii)    a dummy cannot complain if its head or feet are not adequately supported;

(iii)   the weight and posture of a live body will be as at an incident;

(iv)    the manoeuvring of the ladder and patient will have to be more realistically careful, as at an incident;

(v)     it will be possible to deal with the 'casualty' as described in Book 12, Chapter 6, before 'removal'.

### (2) Improvising a step ladder

Occasionally, when firemen are working inside a building, they need to reach the centre of a high ceiling or roof. This is difficult with a conventional pitch. A solution to the problem is to make the two sections of a ladder into a step-ladder. The ladders are extended horizontally on the floor to the required length, one being extended a round further than the other to allow them to interlock. They are then lashed securely at the head and spread at the heel. The spread should not be more than 2.5 m for nine metre and 10.5 m ladders. It may be possible, according to type, to use one section of a short extension ladder as a fixed spreader for a nine metre or 10.5 m ladder. The section is passed through the bottom rounds at the base of the spread ladders and the three sections lashed together (Fig. 2.4 (centre)). If a line is used as a spreader it should be attached by a clove hitch to about the third round on each ladder (Fig. 2.4 (left)). The ladders are then raised with one man footing each.

To make a short step ladder firemen should use the sections of a short extension ladder lashed in a similar way, ensuring the spread is not more than 1 m. (Fig. 2.4 (right)). Usually one man is sufficient for footing.

Fig. 2.4  Centre: constructing a step ladder using two 10.5 m ladders using a rope as a stretcher. Left: using a section of a short extension ladder. Right: a step ladder made from two sections of a short extension ladder.

### (3) Forming a dam

Circumstances sometimes arise where firemen need a large open container for water, e.g. to immerse items needing to be kept cool or during decontamination (see Book 12, Part 2). They can then use sections of extension ladders to form a dam. They should lay three or four horizontally on one string and lash them together into a square or triangle. They should then cover this framework with a salvage sheet, holding it in place by a running bowline passing round the outside of the dam at centre height. When the sheet has settled and the water has reached the required level they should draw the line tight and secure the free end to the standing part with a rolling hitch.

## 5   Using other ladders

### a. Roof ladders

The *Fire Service Drill Book* lays down no drill for the use of roof ladders but there is a note on their use. Firemen should study this carefully.

### b. Hook ladders

This ladder is wholly constructed to withstand a considerable load in line with it. It depends for its stability and safety in use on weight being applied below its point of suspension i.e. the hook. This causes the teeth on the underside of the hook to grip whatever it is resting on and prevent slipping. Firemen should use the ladder only in this way. When scaling a building either from the ground or the head of another ladder they should take care that:

(i)     the toe pieces do not rest on a projection and take part of the weight of the ladder;

(ii)    two ladders do not over-ride.

Only during a test or in an emergency should there be the weight of more than one man on a ladder. Firemen have carried out many rescues by assisting people down this ladder from positions inaccessible to other ladders.

When raising a hook ladder from the ground, a fireman must grasp the third and sixth rounds with the hook on the ground pointing away from his body. He should raise the heel of the ladder slightly and with a sharp cant bring up the head to where the ladder is vertical, balanced and with its hook facing away (Fig. 2.5).

Firemen should study the *Fire Service Drill Book* notes on hook ladders and the hook ladder drill itself with care.

### (1) Hook ladder belt

As already noted, a fireman should always wear a hook ladder belt when using a hook ladder. He should allow it reasonably free

Fig. 2.5 Correct method of picking up a hook ladder to put into a window.

movement round his body. He can then swing the hook round as necessary for hooking on left, right or in front. Thus, for example, on reaching the top of a ladder he can hook on and clear a way for himself before actually mounting the sill.

## 6  Maintenance

Firemen should test to the highest standards all their ladders, in the manner and at the intervals prescribed by the *Fire Service Drill Book*.

### a. Wooden ladders

Wooden ladders should not be stored for long periods in one position as they tend to warp. Firemen should turn them round regularly as well as use them frequently in drills. They should check carefully for any splintering. If it is not serious they should sandpaper the area smooth and varnish it. If however they have any doubts about the ladder they should send it to the workshops for a thorough overhaul and repair. Joints must be close-fitting and resistant to water. Firemen can help guarantee this by making sure the wood is thoroughly varnished.

## b. Metal ladders

Apart from cleaning, metal ladders need little attention other than the prescribed tests and inspections. Most Brigades prefer workshop staff to carry out any greasing necessary. Although warping does not occur, a regular rota of use is desirable, if only to highlight defects as they occur.

# Part 2
# Escapes

## Introduction

In the early C19th the only ladders available to rescue people from fires were of a basic strings and rounds design and, individually, seldom exceeded 5.5 m in length. To increase their manoeuvrability they were mounted on wheeled carriages. An attachment was subsequently added at right angles to the heel of the ladder to improve leverage during elevation.

These appliances were the earliest escapes. Originally, they could only be extended by the addition of extra lengths. Later however a sliding extension worked by wire ropes on drums operated by winch handles was introduced. The addition of a sliding carriage to enable use at any elevation and make handling safer and easier completed the development of the escape.

This appliance remained for several decades a major Fire Service rescue ladder. It was never used widely outside Britain however, the operational need in other countries being met by turntable ladders (see Part 3) and long extension ladders (see Part 1). Now, as noted in Part 1, extension ladders have largely superseded escapes in Britain too. A number are however still in use, and this Part briefly describes their general characteristics and broad principles of working with them.

# Chapter 3
# Escapes

## 1   Characteristics of an escape

### a. Specification

The Joint Committee on the Design and Development of Appliances and Equipment of the Central Fire Brigades Advisory Council has laid down a specification (JCDD 5) for the design and performance of wheeled escapes for Fire Brigade use. This does not specify the material for the escape, which can therefore be of metal, timber with solid strings or timber with laminated strings, nor does it insist on the provision of plumbing gear (see Section 2c below). It does however demand a length of at least 15.2 m, an automatic friction brake, ratchet and pawl with suitable guards for each winch, and automatic pawls at the heel of each extending ladder section. (Some Brigades do nevertheless use escapes without pawls on each ladder section.) It also requires that two men should be able to slip the ladders easily. This Chapter describes the general characteristics of an escape; firemen wanting more detailed information should consult the specification itself.

### b. General appearance

The design of escapes varies from make to make; (Fig. 3.1) shows the essential features. An escape consists basically of a main ladder (1), a middle extension (2) and an upper extension (3) mounted on a sliding carriage and extended by wire rope cables. The middle extension slides within the strings of the main ladder, the upper extension within the strings of the middle. Rollers at the heads of the main ladder and middle extension support them and facilitate their extension. Guide brackets hold in the heels of the extending sections and ladder pawls, when fitted, align the rounds and take the weight from the extending cables when the ladder is pitched. At the heel of the main ladder are the levers. These consist of two lever arms (4) each secured to the base of one string of the main ladder and tied together by cross-members. The levers are usually hinged where they join the main ladder. Lever stays hold the levers at right angles to the main ladder when in the operating position. There may be guy wires from the top of each lever arm to the head of each main ladder string to give support during extension. There are two small lever wheels at the base of the levers to assist in manoeuvring.

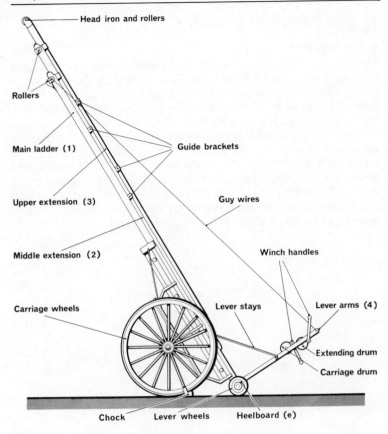

Fig. 3.1 Sketch showing principal parts of an escape.

Also at the heel of the main ladder there may be a heelboard to give added strength and provide a base on which the pulleys for the extending and carriage cables can be mounted. The drums on which these cables are wound are usually between the levers, one carrying the extending cable, the other the carriage cable. Both drums are operated by winch handles and are fitted with safety pawls and/or friction brakes. On some escapes one of the drums is mounted under the main ladder.

The escape carries chocks to put fore and aft to prevent movement of the carriage wheels when it is unshipped. They are usually secured to the carriage by chains. To enable the ladder to move freely when touching the face of a building there are two small wheels on an axle at the head of the upper extension (the head iron and rollers).

23

## 2   Features of an escape

### a. Carriage gear

The carriage wheels are mounted on an undercarriage on which there are guides in which the main ladder slides (Fig. 3.2). Carriage winch gear connected to a cable attached to the head of the main ladder controls the position of the ladders. The rope bends round a pulley on the undercarriage, back via a pulley (1) on the heelboard, to a drum (2) on the levers which is rotated by a winch handle (3). Winding the winch gear elevates the ladders, unwinding causes them to run back under their own weight to their lowest elevation. The backward movement is limited by stops on the carriage or by the setting of the heel chain (4). One end of this clips to a hook on the carriage frame, the other to a hook on the heelboard. Clipping intermediate links to one of the hooks shortens its length as necessary.

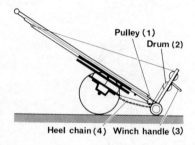

Fig. 3.2  Sketch showing operation of the carriage gear

### b. Extending gear

Fig. 3.3 shows how an escape extends. A cable (1) leads under the main ladder, over a pulley (2) on the underside of the top round and back to a hook (3) in the centre of the bottom round of the middle extension. A second cable (4) leads from a hook (5) in the centre of the same top round via a pulley (6) at the head of the middle extension to a hook (7) in the centre of the bottom round of the upper extension. Winding the cable in on its drum hauls up the main extension, the head of which thrusts against the second cable, thus raising the upper extension. The action is the same when twin cables are employed. A stop is usually bolted onto the main cable at an appropriate distance from the heel of the middle extension. This prevents over-extension of the ladders.

### c. Plumbing gear

When a ladder is pitched on sloping ground its head may be so far off centre as to make it dangerous. To keep it vertical, an escape may

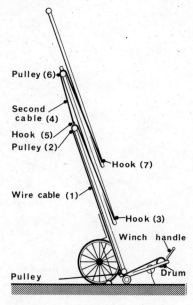

Fig. 3.3 Operation of the extending gear.

have plumbing gear to adjust its angle. Essentially, this allows the main ladder to pivot on either the under-carriage or the carriage wheel axle to about seven degrees either side of the vertical.

### d. Safety devices

(1) Positive pawls

Positive pawls are used on the extending and sliding carriage winches. A positive pawl (Fig. 3.4 (left)) is a pawl which engages with a toothed wheel or ratchet secured directly to a shaft. When the pawl is engaged the shaft can rotate freely in one direction but not in the other. The pawl is engaged when the ladders are being extended or heaved up in the carriage and lifted when the ladder is being lowered or let out in the carriage. When the pawl is lifted, those manning the winch handles take the weight of the ladders and must control the speed of the operation.

(2) Friction brakes

A friction brake (Fig. 3.4 (right)) is similar to a positive pawl. The ratchet wheel is however not secured directly to the shaft but to a drum containing the brake. In one direction the pawl rides over the ratchet teeth, in the other and pawl engages the friction drum which brakes the shaft. Rotation is still possible but it is necessary to wind

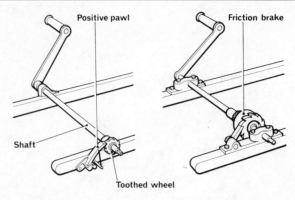

Fig. 3.4 Diagram showing positive pawl (left) and a friction brake (right)

back the winch handle to overcome the braking effect of the drum. The brake takes the weight of the ladder.

(3) Ladder pawls

Automatic pawls are sometimes fitted to the heel of the first extension or to the heel of each extension. They free the extending cables of load when the ladders are extended. To rest the ladders on the pawls, they are lowered until the pawls engage with a round on the section below. A slight extension of the ladder releases the pawls.

# 3   Escape mountings

An escape is mounted on an appliance by a fitting on the rear of the chassis, which supports the axle of the ladder, and by a rest with securing gear on the cab roof, for the heads of the ladders.

Most appliances which still carry escapes have a trunnion push-in mounting. Firemen push the escape into position so that it engages with the steel bar trunnion at the rear of the chassis, then by lifting the heel of the escape to make it head-heavy they use their weight as a counterbalance to lower the head gently onto the rest. A forward push then links the head to the rest, thus removing the need for separate, elaborate, head-securing gear. (Fig. 3.5). To slip the ladder there is a special release handle; when this has been pulled the ladder is slid back and counterbalanced off.

There are several minor variations of this mounting. They usually require special head-securing gear. One particular type still used is the 'Miles' (see Fig. 3.6). With this, two quadrants on the carriage frame roll into two channels on the appliance rear chassis. A lug on each quadrant fits under a bracket and the head is secured by clamping jaws which can be quickly released from inside the crew cab.

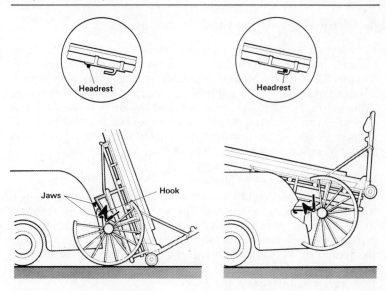

Fig. 3.5 The arrangement of a "push-in" mounting with fixed trunnion.

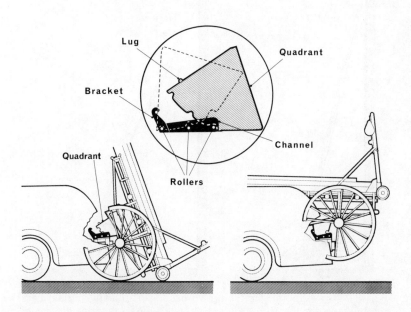

Fig. 3.6 The Miles mounting.

# 4 Working with escapes

Firemen use escapes principally for rescue work but they are also valuable for other purposes, such as providing an external staircase. Chapter 2 mentions a number of points on the handling, pitching and climbing of ladders which apply also to escapes. The notes on escape drills in the *Fire Service Drill Book* and the drills themselves give further basic information. In addition to bearing this in mind, firemen should of course ensure they are familiar with the particular model of escape they are required to use.

### a. Using an escape at an incident

It is sometimes difficult to aim an escape correctly when wheeling it into position against a building. The best method is to get the main axle of the carriage in line with the required window at right angles to it then turn the escape towards it. Once the escape is in position with the weight of its head supported, there is a tendency for it to run backwards. Firemen should guard against this by steadying it with their hands as they insert the chocks.

In positioning an escape, firemen should take care to pitch it neither too far in to a building nor at too low an angle, so that both carriage and lever wheels remain on the ground (see Fig. 3.7). They

Fig. 3.7 The effects of pitching an escape at unsafe angles. Left: pitched too far into the building showing carriage wheels clear of ground. Right: pitched too low showing lever wheels clear of ground.

should use the carriage gear as necessary to ensure this. Carriage wheels must never come off the ground; lever wheels should only do so when an escape is being used for bridging. In the latter case, as many men as possible should man the levers.

If an escape has been pitched into a building, its head must be fully withdrawn before it is lowered, otherwise the extending cable may become slack and the extension eventually come down at a run with a consequent risk of damage or injury.

The use of hook ladders from an escape and the effecting of rescues with an escape are described in the *Fire Service Drill Book*.

### b.  Moving an escape

Firemen can safely wheel a housed escape in a vertical position on its carriage and lever wheels. They must however lift the lever wheels before executing an acute turn and not drag them sideways.

### c.  Storage an escape

When an escape is unlikely to be used for some time firemen should fold its levers and prop it against a wall with its chocks down, so that it occupies the least space.

### d.  Maintenance of escapes

An escape should be examined, cleaned and greased regularly, with particular attention to the condition of extending and carriage cables. Standard tests should be carried out as prescribed by the *Fire Service Drill Book*.

# Part 3
# Turntable ladders

## Introduction

As buildings increased in height and business areas became more congested Brigades found that the escapes, described in Part 2, were often inadequate. They were too short and could not be used as water towers. To overcome these difficulties by modifications to the escapes would have meant adding elaborate trussing which would have made them too heavy and unwieldy.

A new ladder was therefore necessary. Clearly, to meet the requirements it would be too heavy for manhandling, so the basis for development had to be an extending ladder permanently mounted on a mobile chassis. Early attempts gave the extra ladder-length and sufficient stability for use as a water tower but had other drawbacks: the appliances were too long for easy manoeuvring, the ladders could only be elevated at one angle, and they could not be rotated. The overcoming of these difficulties led to the basic type of modern turntable ladder. Further developments in such areas as details of design, construction materials, means of operation (mechanical and hydraulic) and the addition of extra features such as ladder cages followed.

This Part of Book 5 looks at the general characteristics of turntable ladders and broad principles of working with them, and it refers to the principal types currently in use in Great Britain.

# Chapter 4
# General information on turntable ladders

## 1.  Terminology

Firemen can carry out a number of different manoeuvres with a
turntable ladder and use it for a variety of purposes. To avoid any
confusion the Fire Service has adopted a standard terminology for
use when referring to the various t.l. operations.

| | |
|---|---|
| Depress | to lower the head of the ladder by reducing the angle of elevation. |
| Elevate | to raise the head of the ladder by increasing the angle of elevation. |
| Extend | to increase the length of the ladder |
| House | to reduce the length of the ladder |
| Plumb (to right or left) | to keep the centre line of the ladder in a vertical plane by eliminating any tilt to one side when the ladder is extended on a slope. This increases stability and obviates side stress. |
| Projection | the horizontal distance measured from a vertical line dropped from the head of the ladder to the rim of the turntable. |
| Shoot up | to extend the ladder with a fireman already at its head |
| Train (to right or left) | to move the head of a ladder by rotating the turntable. (N.B. manufacturers tend to use the expression 'slewing'.) |

Firemen should first familiarise themselves with these terms and with
the various parts of a t.l. named in Fig. 4.1.

## 2   Design of a turntable ladder

### a. Specifications

The Joint Committee on the Design and Development of Appliances
and Equipment of the Central Fire Brigades Advisory Council has

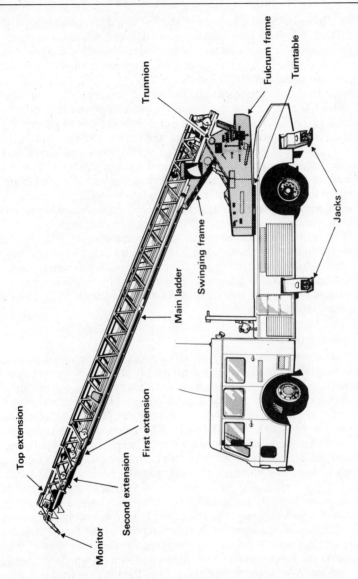

Fig. 4.1 The principal features of a turntable ladder.

laid down specifications for the design, construction and performance of t.ls. The specification currently applicable is JCDD 36 published in June 1976 and amended in December 1980. It replaces the earlier JCDD 15, also on hydraulically operated ladders. (This superseded JCDD 14 on mechanically operated ladders, which were no longer being made.) This Chapter describes only some of the

33

more important general features shared by all ladders complying with JCDD 36. Firemen who wish for more detailed information on the requirements such ladders should meet should consult the specification itself. Chapters 5-8 gives a more detailed descriptive account of some particular ladders.

Firemen should be aware that there are at present (1983) a number of older t.ls still on the run which do not comply with JCDD 36. JCDD 15 laid down similar but not such exacting requirements. Among more noticeable differences are that it did not specify the provision of a cage or the ability to work below the horizontal. Firemen should note that at the time of writing JCDD 15 is being revised and reinstated to cater for Brigades which do not require cages on their t.l.s. Mechanically operated ladders differed primarily in their means of operation; they were less sophisticated, had less delicate controls and performed less smooth movements.

The requirements set out by JCDD 36 are minima or maxima only. Manufacturers may choose to exceed them and Brigades may always lay down stricter or additional requirements, perhaps to meet special conditions. Turntable ladders must also conform with current *Motor Vehicle (Construction and Use) Regulations* and with *Road Vehicle (Lighting) Regulations*.

## b. General description

In essence, a turntable ladder is a self-supporting and power-operated extension ladder mounted on a turntable. Early ladders were of wood strengthened by angle irons and steel trussing but an all metal construction has now superseded this design. The ladder assembly is mounted at the rear of a heavy, self-propelled, chassis approximately above the back axle (see Fig. 4.1). The chassis design takes into account that the appliance usually stands laden. Its wheel base does not exceed 5.3 m nor its over-all width 2.5 m; except at the axle casing it has at least 230 mm ground clearance. At the base of the ladder assembly is a turntable which rotates in a circular track bolted to the chassis. On this is mounted the fulcrum frame, a strong steel structure which supports the rest of the ladder assembly and houses the operating mechanism. At the fulcrum point of this frame there is a trunnion on which pivots the swinging or elevating frame. This supports the ladder sections and forms a mount on which to operate the ladder. Also fitted to the nearside of the fulcrum frame or on a separate console are control levers for the ladder. There may also be a secondary set of controls and indicators in the ladder cage, or capable of being fitted at the head of the ladder, but the main controls at the base will at all times be capable of over-riding them. All t.ls have plumbing gear which keeps the ladder plumb on gradients up to about seven degrees.

The ladder itself usually consists of a main ladder, secured by a strong pivot bearing to the swinging frame, and three or four extensions which extend telescopically. There is usually also a small ladder to facilitate access from the ground. The extensions are each extended, and housed in some cases, by separate flexible steel cables. The method is similar to that used with escapes (see Chapter 3), but t.1s are of course power operated. At the heel of each extension there are usually pairs of ladder pawls which enable the extensions to rest upon each other with positive safety independent of the cables. On other ladders the 'pawl' action is achieved by the operation of special valves inside the extending rams.

When fully extended at maximum elevation a t.1. should reach a height of 30.5 to 33.5 m above the ground. Though longer ladders can be constructed none would be long enough for the tallest modern buildings. The JCDD specification therefore does not cater for them. It is the practice in the UK for buildings over 24.4 m to have internal firefighting facilities and means of escape.

### c. Other features of appearance

(1) Cages

All ladders complying with JCDD 36 have a cage for the head of the ladder. This may be either a fixed or demountable light rescue cage capable of carrying at least two people or a permanently attached rescue cage (working platform) capable of carrying at least four (see Plate 7).

(2) Monitors

At the head of a ladder or front of its cage there is a monitor (see Fig. 4.2). A man at the head of the ladder normally operates this, but it

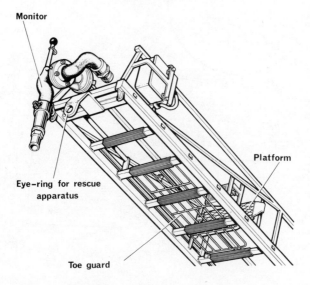

Fig. 4.2 Typical turntable ladder monitor

35

may also be possible to operate it from the ground by lines. The monitor is fed by a special length of hose and a fixed lightweight water supply pipe may be provided to form a permanent connection between the monitor and the top ladder section.

### (3) Operator's platform etc.

If the ladder does not have a cage permanently fitted, it will have secured near its head a hinged platform for the monitor operator. Unless the ladder design makes it unnecessary, the platform will have hand-rails and a toe guard. There will also be a steel ring running on a vertical bar to which the operator can secure himself with a belt and snap hook as a safety measure.

### (4) Rescue gear

All ladders without a permanently fitted cage must have provision for lowering by line in order to effect a rescue (see Fig. 4.3). The line used is about 70 m long, fitted with a sling, and made up on a cradle or

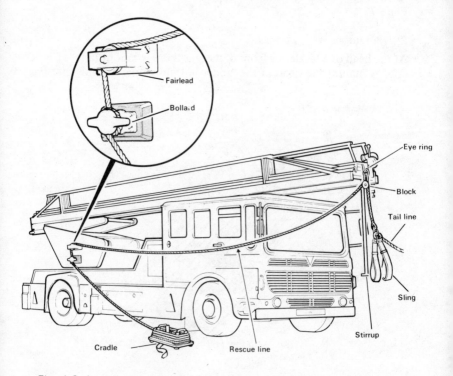

Fig. 4.3 An example of a rescue apparatus on a turntable ladder.

trolley. To take this line the ladder has at its head, in the centre, either a permanently fitted pulley or an eye to which a pulley can be attached by a snap-hook. For control of this line the turntable or turret of the appliance has a fair-lead and a bollard.

## 3    Safety devices

### a. Automatic safety devices

All turntable ladders have automatic safety devices. These includes:

| | | |
|---|---|---|
| (i) | limiting cut-outs and stops | to slow down and stop ladder movements at maximum elevation, extension and depression below the horizontal; and when fully housed; |
| (ii) | intermediate or safety stops | to stop extension or depression of the ladder when the safe operating limits are reached; |
| (iii) | impact stops | to prevent damage to the ladder if it strikes an obstruction; |
| (iv) | control interlocks | to prevent any ladder movement until the jacks are down and any retraction of the jacks until the ladder is housed and depressed on the head-rest; |
| (v) | plumbing mechanism | see Section 2b. above. There are stops at the extreme limits of plumbing; |
| (vi) | depression and extension safety devices | where elevation is by hydraulic rams the rams have locks and, if the rams are twin, one will be capable of supporting the ladder by itself. Extension mechanism is duplicated so that if one half fails the other will still hold the ladder. |

### b. Jacks and axle locks

A turntable ladder needs a solid working base. The varying loads it imposes in use must therefore not be allowed to compress the road springs and tyres. A total of four jacks fitted in front of, and behind, each rear wheel help achieve this by being lowered to the ground and taking the excess weight of the ladder and chassis off the tyres. They are so designed that it is impossible for them to retract even when continuously loaded. An axle locking device prevents the chassis rising on the road springs when a ladder, extended on the opposite side, reduces the loading. It may consist of a claw or eye on each side of the chassis, hinged to engage with the rear axle, which it locks to the chassis. The design of some modern t.ls removes the need for axle locks. T. ls also carry chocks for the road wheels.

#### c. Alignment of rounds

Where the different extensions of a ladder over-lap the rounds of each may not be in line with those of the others. This could be inconvenient or dangerous for a person climbing the ladder and there is therefore always a means of aligning the rounds, possibly by a separate 'rounds in line' control.

#### d. Communications

All turntable ladders have provision for the fireman at the head of the ladder and the operator to communicate in all circumstances.

## 4   Safe working indicators

### a. Inclinometers

It is most important to maintain the stability of the ladder. Principal factors affecting this are the length of extension, the angle of elevation and the loading of the ladder (see Chapter 10, Section 2a). Each ladder therefore has a device which indicates its elevation and shows the maximum permissible extension at different angles for different loadings. This is known as the inclinometer.

The inclinometer, or a separate field of operations indicator (see plate 8), also gives information on working safely with a ladder when it is being used in bridging, i.e., with its head supported, or as a water tower.

### b. Other indicators

Turntable ladders also have the following safe working indicators.

(i)     safe load indicator; this gives both a visible and an audible warning when the maximum permissible load limit is reached;

(ii)    ladder position indicator;

(iii)   plumbing indicator;

(iv)    'rounds in line' indicator;

(v)     means to show when the ladder sections are fully housed;

(vi)    means to show that the hydraulic power supply system is operating at working pressure.

## 5   Operation and performance

### a. General

A turntable ladder is designed for use as a water tower, in rescues, and for any other purposes which require an aerial platform.

It performs the following operations:

(i)     depression;

(ii)    elevation;

(iii)   extension;

(iv)    housing;

(v)     plumbing;

(vi)    training.

In modern t.1s all the movements except depression are power-operated. Depression, elevation, extension and housing are controlled by levers and are infinitely variable throughout the speed range. A t.1. has a road turning circle of no more than 20.7 m in either direction. It is designed for speed (at least 80.5 k.p.h. on the flat), good road holding and fast cornering, and to give a comfortable ride to the crew. It is capable of use at any angle up to at least 75° but no more than 78° and can also work below the horizontal. A ladder can be trained through 360° to right or left at any angle exceeding seven degrees above the horizontal. If the cage is detachable it will be possible to fit it from ground level with the ladder fully depressed and housed. The t.1. monitor has a range of 10°–15° either side and a vertical range as great as possible and not less than 100°. The minimum pressure of the pump supplying it is seven bar.

### b. Power take-off

A turntable ladder usually has a power take-off which forms part of the road gear box and transmits adequate power to meet the requirements of the ladder mechanism. It is operated by an engagement lever in the driving compartment of the appliance unless it can be engaged without the use of the clutch. In an emergency, it is always possible to operate a t.1. manually or from a secondary power source.

Experiments have been conducted to design a t.1. with an engine separate from the appliance engine to power the ladder. Some models are now beginning to come into the Service.

### c. Pumps

If the t.1. appliance also provides a pump, the pump will comply with JCDD 29 and be a separate self-contained unit. This simplifies transmission drives, allows individual control of pump and ladder and permits easy removal of the unit for maintenance.

### d. Acceptance tests

All turntable ladders complying with JCDD 36 are subject to acceptance tests which are laid down by the specification. The tests may take place at the works, the Brigade, or elsewhere. Certain of them can be omitted if the manufacturer submits a certificate to guarantee that another appliance of the same design and construction has already passed them.

# Chapter 5
# The Magirus DL30 turntable ladder

## 1   General appearance

The ladder set consists of a main ladder and three sections, all of steel, giving a maximum extension of 30 m. The set is specially shaped to expose a minimum surface to the wind and the ladder rounds have a non-slip tread. Design may be modified to suit the particular chassis used or the requirements of the Brigade. The ladder can be elevated to 75° from the chassis and depressed to –15°, below the horizontal. Pawls are unnecessary as the ladder is retracted by means of return cables (see Section 3b below). Principal controls for the ladder are concentrated on a small console where the operator sits on the turntable at the heel of the ladder (see Fig. 5.1).

Amongst other features of the ladder are:

(i)     a rescue cage that can be permanently fitted or detachable and is capable of holding two people;

(ii)    inclined jacks which can be operated from either of two boxes fitted at the rear of the vehicle (see Fig 5.2). They have a simple pressure button which brings in the axle locks first then operates the jacks;

(iii)   separate oil pumps for each of the three main ladder movements;

(iv)    a microphone and loudspeaker at the head of the ladder and by the ground operator's seat;

(v)     a short extension ladder, mounted at the rear of the cab, for gaining access to the heel of the ladder;

(vi)    a monitor, either permanently attached to the rescue cage or mounted at the head of the ladder, which can be trained 15° right or left;

(vii)   a 110 litre hydraulic oil tank;

(viii)  a ring at the head of the main ladder to enable the appliance to be used as a crane

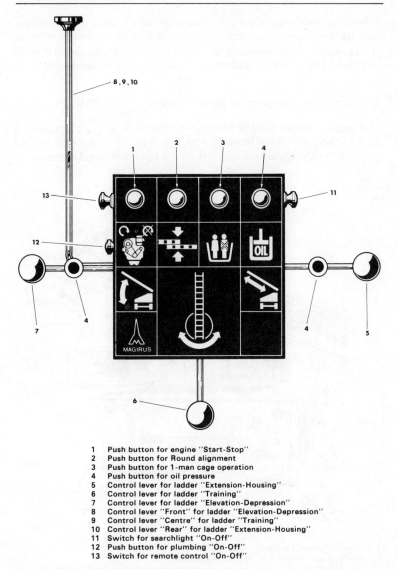

1   Push button for engine "Start-Stop"
2   Push button for Round alignment
3   Push button for 1-man cage operation
4   Push button for oil pressure
5   Control lever for ladder "Extension-Housing"
6   Control lever for ladder "Training"
7   Control lever for ladder "Elevation-Depression"
8   Control lever "Front" for ladder "Elevation-Depression"
9   Control lever "Centre" for ladder "Training"
10  Control lever "Rear" for ladder "Extension-Housing"
11  Switch for searchlight "On-Off"
12  Push button for plumbing "On-Off"
13  Switch for remote control "On-Off"

Fig. 5.1  Diagram of console controls

## 2   Principles of operation

The ladder movements are effected hydraulically by separate pumps driven by a power take off from the main road engine. The separate pumps prevent the slowing down of a movement when another is in progress at the same time and avoid unnecessary oil heating.

Plumbing is automatic but elevation/depression, extension/housing, and training are controlled by levers. All drive elements such as elevating rams and extension motors and cables are duplicated; each can effect the required movement independently if necessary.

## 3 Operations controlled from the rear of the ladder

The main controls for the Magirus turntable ladder are sited at the heel of the ladder (see Plate 10 and Fig. 5.2). They can be operated either from the ground to the left of the operator's seat or from the

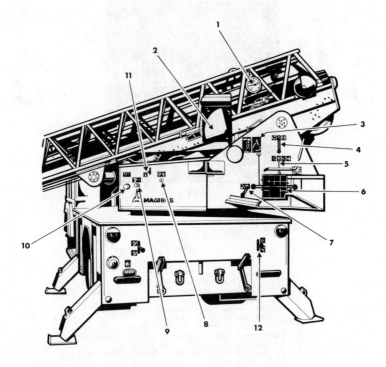

1  Load indicator for lifting device
2  Field of operations indicator with symbols
3  Loading data for ladder used as lifting device and ladder used as water tower
4  Control lever for hand operation "Depression", power operation and hand operation "Elevation"
5  Control lever for hand operation "Extension", power operation, hand operation "Fast housing" and hand operation "Slow housing"
6  Instruction plate with symbols
7  Control lever for "Ladder Operation" and "Lifting Operation"
8  Push-button for "Emergency Operation of Stabilizing Jacks"
9  Connecting pin for hand-crank operation "Training"
10  Hand pump lever connecting pin for hand pump
11  Control lever for hand operation "Plumbing"
12  Control lever for "Extension-Retraction of Stabilizing Jacks"

Fig. 5.2 Sketch showing main controls, jack controls and jacks.

seat on the console itself. From the ground there are three control levers: on the left for elevation/depression, on the right for extension/housing, and in the centre for training. From the console seat there are likewise three levers, in a line, connected with these. The console also has buttons and switches for various other controls.

### a. Elevation/depression

These movements are effected by two hydraulic cylinders both fitted with hydraulically controlled non-return valves which prevent the ladder dropping back.

### b. Extension/housing

The ladder sets extend and house over nylon rollers. There are two cables for extension and two for housing. The housing cables make pawls unnecessary and housing is possible at any angle between maximum depression and maximum elevation. Full extension of the ladder is possible while it is horizontal.

### c. Training

To train the ladder there is a hydraulic motor with a self-locking worm gear connected to the turntable ring gear. The ladder can rotate endessly through 360°.

### d. Other operations

Amongst other switches on the console are one to turn on and off the remote control (see Section 7), one to align the rounds and one to control engine speed. There is a 'dead man's' foot button which must be kept depressed to maintain the oil pressure and keep the ladder operative. The communications system is switched on automatically for the top man to speak to the operator as soon as the jacks are extended. To speak to the top man the operator has to keep a button on the hand microphone pressed.

## 4   Plumbing

Plumbing depends on mercury switches which control two hydraulic rams. The rams adjust the rear frame, complete with swinging frame and ladder until the ladder is plumb (up to seven degrees either side). The whole front part of the frame is joined to the chassis by the turntable ring whilst the rear part and elevating frame can be tilted around a horizontal axis to achieve the plumbing. The ladder is automatically centred when depressed onto the headrest. The plumbing can be switched off if necessary for very fine manoeuvres.

# 5   Safety devices

### a. Axle lock and jacks

The four inclined jacks fitted fore and aft of the rear wheels provide a firm working base and eliminate movement of the chassis frame and tyre deflection. A hydraulic interlock prevents the jacks extending before the axle locks operate and ensures that, until there is a positive pressure between the ground and jacks, the ladders cannot be lifted from the headrest. Once the ladders are off the headrest it is not possible to retract the jacks.

### b. Limit stops and locking devices

All ladder movements stop automatically as soon as they reach maximum elevation, depression and extension or are fully housed. When approaching these positions the movements slow down so that no damage occurs as the actual limits are reached. On sloping ground the field of operations indicator (see Section 6b below) will prevent elevation above 75° and depression below –15°; the moving pointer touches the contact plate for the one man free — standing limit and the electric contact prevents further ladder movement.

### c. Safe load indicator

All forces acting on the ladder — weight, wind pressure, guy lines, etc. — are measured electronically. If there is an excess load the system first cuts out then allows only depression and housing.

### d. Emergency device

Under the console seat is a lever for the emergency device. The operator can use this to return the ladder from one of its cut-out limits *if it is not possible to do so by normal means.*

### e. Automatic lever return

If the oil pressure fails all control levers return to neutral, to prevent uncontrolled movement when the pressure is restored.

### f. Bridging cut-outs

Bridging cut-outs are located at the head of the second extension, under the strings of the top extension. If the ladder is over-loaded, subjected to high winds having the same effect, twisted, or depressed while the head is resting down, these cause the levers to return to neutral. Only elevation and training are then possible until the ladder is in a safe position.

### g. Dead man's button

see Section 3d above.

# 6   Safety indicators

### a. Safe load indicator

In addition to stopping ladder movements the safe load indicator also operates an audio and a visual signal when the limit is exceeded.

### b. Field of operations indicator

The indicator is suspended like a pendulum and shows the angle of the ladder to the horizontal (see Fig. 5.3). A moving pointer indicates the position of the ladder with regard to extension, height, angle, etc. before it reaches its limits. Symbols along the top of the indicator are illuminated to give a further visual warning when limits are reached. Other lights indicate that the oil pressure is on, that the plumbing is on, and that the rounds are aligned.

### c. Warning light

A light in the appliance cab operates when the jacks move from the fully housed position.

# 7   Remote controls

Where a rescue cage is fitted or carried there are facilities for remote control of the ladder movements from the cage (see Fig 5.4). These duplicate most of the ground controls but there is no full field of operations display and control over movements is combined in one lever. If required there can also be a remote control box with some 15 m of cable. This enables the operator to manoeuvre the ladder while standing away from it. On modern models the box can also be plugged into the top of the ladder to give the man at the head a measure of control.

# 8   Operational range of the Magirus turntable ladder

Fig. 5.5 shows the general range of a Magirus t.1. Firemen should bear in mind during operations the points brought out above and dealt with in Chapter 10. In windy weather guy lines must be used at wind speeds of Beaufort Scale Force 5 (25 k.p.h.) and above and the ladder must be partially housed at wind speeds of Force 7 (55 k.p.h.) and above. When the ladder is in use unsupported as a water tower, the following table of exact nozzle sizes, pressures and output should not be exceeded:

| Nozzle diameter mm | Monitor pressure bar/cm$^2$ | Output 1/min. |
|---|---|---|
| 26 | 8.0 | 1260 |
| 30 | 7.0 | 1560 |
| 32 | 6.0 | 1640 |

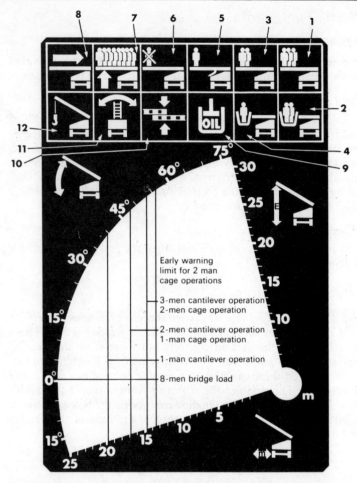

Fig. 5.3 Sketch of field of operations indicator.

Symbol lights up if:
1    No more than 3 men are admissible at ladder point
2    No more than 2 men are admissible in the cage
3    No more than 2 men are admissible at ladder point
4    No more than 1 man is admissible in the cage
5    No more than 1 man is admissible at ladder point
6    Load in free-standing position is prohibited
7    With the ladder point placed 8 man bridge load is admitted
8    Ladder must be housed
9    Oil pressure is engaged
10   Rounds of ladder are aligned
11   Plumbing is disengaged
12   Lifting operation is engaged

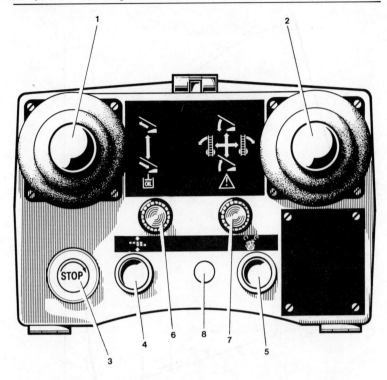

1   Control lever for ladder "Extension-housing"
2   Control lever for ladder "Elevation, Depression, and Training"
3   Push-button with locking mechanism for cutting system in and/or
     instantly cutting it out (Stop-button with locking mechanism)
4   Push-button for Round Alignment
5   Push-button for engine "Start-Stop"
6   Pilot lamp for "Oil Pressure On"
7   Pilot lamp for "Early Warning"
8   Panel illumination switch

Fig. 5.4 Diagram of cage controls

The ladder should not be elevated above 70° and there should be no
projection beyond 14 m. In bridging, the ladder can be loaded with
eight persons equally distributed. The white symbol '8 men bridging
load' on the field of operations indicator will light up.

Access to the ladder is by use of the short extension ladder supplied
(see Section 1) or by depressing the ladder head to the ground. The
latter manoeuvre is however usual only for entering or leaving the
cage.

47

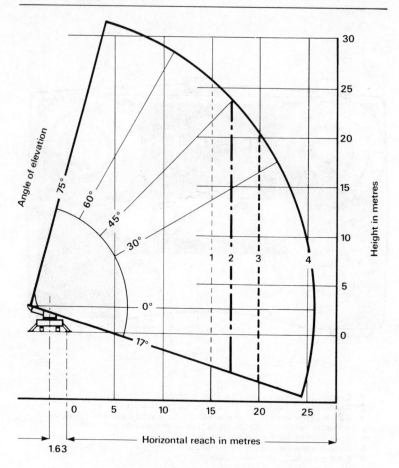

| 1 | Top of ladder  3 men (255kg) |
|   | Rescue cage    2 men (170kg) |
| 2 | Top of ladder  2 men (170kg) |
|   | Rescue cage    1 man (100kg) |
| 3 | Top of ladder  1 man (100kg) |
| 4 | Top of ladder supported |
|   | 8 men (bridge load) |

Fig. 5.5 Diagram of operational range

# 9   Hand operations

In the event of an engine failure it is possible to depress, house and plumb the ladder and to retract the jacks by means of a manual hydraulic pump. The operator must select the appropriate circuit from the levers on the side of the ladder frame before pumping starts. Housing is possible at either fast or slow speed. Training is carried out by means of a cranked handle fitted to a square pin on the ladder frame (see Plate 11).

# Chapter 6
# The Magirus DLK 23-12 turntable ladder

The preceding Chapter has described the Magirus DL30 turntable ladder. These appliances have been widely used by the British Fire Service and many are still operational. Recently, however, the manufacturers have introduced a new model, the DLK 23-12 (see Plate 12). There are two versions: one of standard design, the other of a special low-line construction. Both differ somewhat from the DL30 and this Chapter summarises the major differences.

## 1   General appearance and operational capabilities

The following table sets out the exact principal dimensions and operational abilities of the ladder:

|  | Standard | Low-line |
|---|---|---|
| Over-all length | 9.6m | 9.6m |
| Over-all height | 3.25 | 2.85m |
| Over-all width | 2.5m | 2.35m |
| Over-all width with jacks in use | 2.5–4.5m | 2.35–4.5m |
| Minimum extension | 30m | 30m |
| Minimum elevation | 75° | 75° |
| Maximum depression | –12° | –12° |

Fig. 6.1 shows the operational range of the ladder.

The basic construction of the ladder remains as before and it shares a number of the features of the DL30. Particularly in the low-line model, however, the appearance is different. In both standard and low-line models the ladder set itself has been made wider and more rigid. It has a completely unobstructed climbing field. A stretcher can be mounted at the cage.

### a. Jacks

The four jacks can be operated independently and are infinitely adjustable within the stated width limits (i.e. 2.5–4.5m on the standard model, 2.35–4.5m on the low-line model). At their minimum extension they remain within the vehicle outlines (see Fig. 6.2), thus allowing operation within more confined spaces; the wide maximum extension allows an increased operational range. Unlike

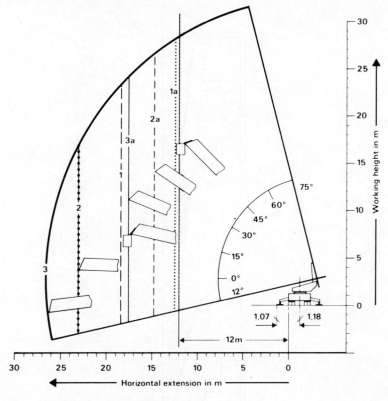

Fig. 6.1 Diagram showing operational range

the DL30 jacks lie flat to the ground (see Plate 13). Each is fitted with a warning light. During operations the rear wheels remain firmly on the ground. The ground pressure is controlled by sensors.

## 2   Operation of the ladder

The DLK 23-12 operates on the same basic principles as the DL30, but there have been some changes in detail. The main controls remain at a redesigned operator's central control station, fitted with a seat, on the turntable at the rear of the ladder. They consist principally of two levers: one for extension/housing, one for elevation/ depression and training. The ladders extend and house over plastic blocks and rollers; the twin extending cables are laterally arranged. An infinitely adjustable extension control processes all values and calculates the maximum ladder extension permissible in relation to the width of the jacks. Communications between the operator's console and ladder

51

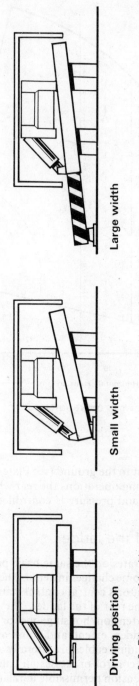

**Driving position**        **Small width**        **Large width**

Fig. 6.2 Diagram of jacking range.

head are now controlled by a foot operated switch at the console. There is a button to turn the engine on and off and a foot-operated switch to select one of two possible engine revolutions.

## 3   Plumbing

The plumbing of the ladder, which is effective up to seven degrees is automatic. It operates on a system of live rings which level the complete turntable and ladder set when the ladder is raised from the headrest. The main ring then remains level, keeping the ladder plumb throughout operations without further movement of the system. The ladder is automatically centred when depressed back onto the headrest.

## 4   Safety features

The DLK 23-12 shares the safety features of the DL30. Its field of operations indicator has been slightly re-designed and takes into account the effect of the different jacking positions possible. The ladder's limits change according to the extension permissible and all ladder movements stop automatically when the limits are reached. The illuminated symbols include an indication that jacking has not been completed. There are safety devices to prevent uncontrolled movement of the ladders in case of a fracture in the oil pipe system.

# Chapter 7
# The Merryweather turntable ladder
# ('C' type)

## 1   General appearance

The Merryweather consists of a main ladder and three extensions composed wholly of welded steel. Their total length when fully extended is 30 m and the square section rounds which are fitted with non-slip rubber treads are 300 mm part. Elevation of up to 75° from the chassis is possible. The ladders have rectangular steel trussing and pawls near the heel of all extending sections. These are fully automatic and pawling can be achieved at 900 mm intervals. Fig. 7.1 shows their operation. The ladders have the following features:

(i)    at the head of the main ladder, a ring to enable use as a crane;

(ii)   rings for the attachment of guy lines and a rescue line;

(iii)  a folding platform for the man at the head of the ladder;

(iv)   a monitor with lateral movement of 15° either side and a vertical range of 100°;

(v)    mounted on the trussing, a combined telephone transmitter and receiver.

The controls and indicators for operation of the ladders are grouped together on a console fitted with a seat for the operator. This console stands on a platform fitted to the near side of the fulcrum frame so that as the operator trains the ladder round he is carried with it and remains in the most convenient position facing the head of the ladder. The turntable for the fulcrum frame is mounted on sealed double row ball bearings.

Other features of the appliance are:

(i)    jacks          the type varies according to the model: they may screw straight down or may swing out into the operating position and be locked into place by spring-loaded plungers. They are fitted with ball-jointed feet;

(ii)   axle locks      these are operated by foot pedals at the rear of the appliance. They consist of hinged levers with rings at the end which engaged with hooks attached to the rear axle covering.

54

(iii)  a hydraulic   this is located in the chassis and driven by the
       oil pump      road engine through a power take-off engaged
                     from the driver's cab;

(iv)   a 225 litre
       oil tank      also in the chassis.

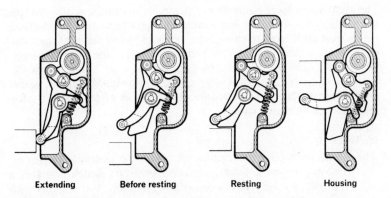

| Extending | Before resting | Resting | Housing |

Fig. 7.1 Automatic ladder pawl showing working positions.

## 2  Principles of operation

All the ladders movements are effected hydraulically. In the fulcrum
frame there are four main hydraulic circuits; for elevation/
depression; for extension/housing; for training; and for plumbing.
The oil from the tank is fed to them by the pump through a rotary
junction in the centre of the turntable at a pressure of 70 bar. A
control unit keeps this pressure constant, irrespective of demand, so
that the speed of any ladder movement depends entirely on the
position of the control lever (see Section 3 below). The control
automatically governs engine speed as the demand for fluid increases
or decreases.

Before operations it is necessary to pressurise the system by
depressing a foot pedal at the base of the control console. This is a
precaution against inadvertent movement of the ladders.
Elevation/depression, extension/housing and training are then all
controlled by levers on the console; plumbing is automatic,
Hydraulic rams effect elevation; Hydraulic motors the other
movements.

## 3  Operations controlled from the console

Fig. 7.2 shows a Merryweather console. There are three levers. Those
for elevation/depression and extension/housing are hand operated;

that for training can be operated either by hand or by knee. When first moved the levers encounter light spring pressure and the speed is slow; as a lever is moved further, so the speed increases. Moving a lever in any particular direction results in a corresponding movement of the ladders.

### a. Extension

The right hand lever controls extension. A hydraulic motor on the turntable drives the extension winch through a reduction gearbox. A single steel cable from the winch drum passes over a pulley at the head of the main ladder and is anchored to the heel of the first extension. Other extensions are operated by twin cables. They house by gravity but with the lever in neutral will only do so slowly. Movement of the lever towards the operator reduces the braking effect of the motor and thereby increases speed.

### b. Training

The centre lever controls training. A reversible hydraulic motor on the turntable drives a worm reduction gear-box which operates a pinion projecting through the box of the turntable; this engages with an internal ring gear attached to the chassis.

### c. Elevation

The left hand lever controls elevation. Twin single acting rams controlled by oil flowing into and out of a hydraulic lock safety device at their base effect the movement. Depression is by gravity and the rate of flow of oil from the rams affects its speed. Maximum elevation is 75° to the chassis, minimum nought. In the event of damage one ram alone can support the ladders.

### d. Communications

The combined microphone and loudspeaker are set into the centre top of the console and controlled by the right hand switch and button. There is a buzzer for the top man to attract the operator's attention when the telephone is not switched on.

### e. Remote engine starter

If the engine stalls the operator can re-start it without leaving the console by pressing the left hand button.

## 4   Plumbing

A free swinging plumb bob on the underside of the swinging frame controls plumbing automatically. Its movement operates a spool valve which allows fluid to drive a hydraulic motor in the required direction. This motor rotates a screwed shaft at the rear of the swinging frame by means of a nut running on the shaft. Plumbing is

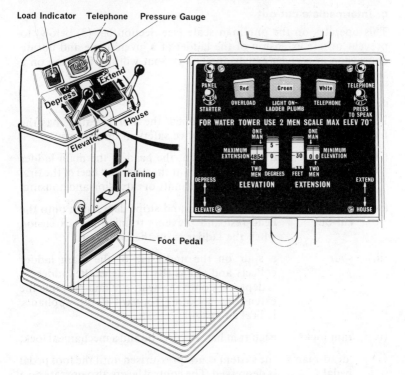

Fig. 7.2 Control console. Right: details of control panel.

limited to seven degrees either side by means of a metal frame surrounding the plumb bob. The front of this frame is so shaped that at angles below 45° the plumbing is automatically centralised.

## 5   Safety devices

### a. Limit stops

The purpose of limit stops is to slow down and stop the ladders at their limits of travel. The Merryweather has them for extension/housing and elevation/depression. The travel of the plumb bob is limited by its frame.

### b. Impact stops

Impact stops prevent damage to the ladder if it touches a solid object during operations. They take the form of pressure limiting valves in the hydraulic feed pipes. The training impact stop is built into the turntable. The impact stops for elevation and extension depend on a pair of cams on the off-side fulcrum frame.

#### c. Intermediate cut-out

This operates on the one man scale (see Section 6a. (3) below) to prevent unsafe extension of the ladder at a given angle and unsafe depression at a given extension. The cut-out will not function on a steep hill.

#### d. Other safety devices

In addition to those already mentioned, the Merryweather has the following features designed to enhance safety in operations:

| | | |
|---|---|---|
| (i) | rubber buffers | these are fitted at the head of the main ladder and contact cups at the head and heel of the first extension at the limits of extension and housing; |
| (ii) | head-rest cut-out | this slows down and stops depression onto the head-rest and prevents training and extension when the ladder is on the head-rest; |
| (iii) | spur | a spur on the underside of the main ladder collects and aligns the rounds as the ladder is depressed onto the head-rest and prevents the extensions shooting forward if the appliance is braked heavily; |
| (iv) | ram locks | each ram has a hydraulic and a mechanical lock; |
| (v) | 'dead-man's' pedal | the system is not pressurised until the foot pedal is depressed. The control levers also operate on a 'dead-man's handle' principle and return to neutral if released; |
| (vi) | plumbing stops | there are stops at either end of the curved track supporting the heel of the main ladder. |

## 6   Safety indicators

In addition to the various safety devices mentioned above, the Merryweather turntable ladder has a number of indicators which either show whether operations are being conducted safely or give information on factors to be taken into account.

#### a. On the console

(1) Inclinometer

This records the angle of elevation relative to the chassis in ten degree intervals up to 30° and five degree intervals thereafter. A floating zero records the angle of elevation relative to the true horizontal, for when the appliance is at work on a slope.

(2) Extension meter

This indicates the extension of the ladder at fixed intervals.

(3) One man and two man scales

To the left of the inclinometer are scales showing the maximum permissible extension for the elevation shown, one with one man at the ladder's head, the other with two. To the right of the extension meter are similar scales showing the minimum permissible elevation for the extension shown.

(4) Over-load indicator and lateral strain alarm

When the ladder is over-loaded the needle of the load indicator dial at the top left of the console (see Fig. 7.2) moves into the red, an alarm bell rings and a red light is lit.

The ladder also has a lateral strain gauge set to sound a buzzer on the console when subjected to a side load equivalent to a wind speed of 40 k.p.h.

(5) Pressure gauge

At the top right of the console is a gauge showing the pressure in the hydraulic system. With the pedal at the base of the console depressed, this should be between 70 and 80 bar.

(6) Ladder central lights

The two centre orange lights on the console indicate that the ladder is in line with the head rest when they are both on. Illumination of either separately indicates that the ladder is to that side.

(7) Ladder plumb light

A green light at the bottom of the panel indicates that the ladder is being plumbed automatically. If it goes out, the plumbing limit has been reached and the ladder is unstable.

### b. Other

(1) Burgee

A burgee is fitted to the head of the ladder to indicate the direction and to a lesser extent the strength of winds affecting it. The swivelling staff is weighted to remain vertical at all times. It must not be used as a hand-hold.

## 7   Operational capabilities

In operations with a Merryweather t.1, firemen should bear in mind the various points mentioned in Chap 10.

Fig. 7.3 shows the operational range of a Merryweather ladder. The maximum distance over which the ladder can be bridged is 15.5 m at right angles to the chassis, less from other positions. When in use as a water tower the ladder's maximum elevation is 70°. Determination of the pump pressure must take into account the pressure loss due to the height above the pump at which the branch is

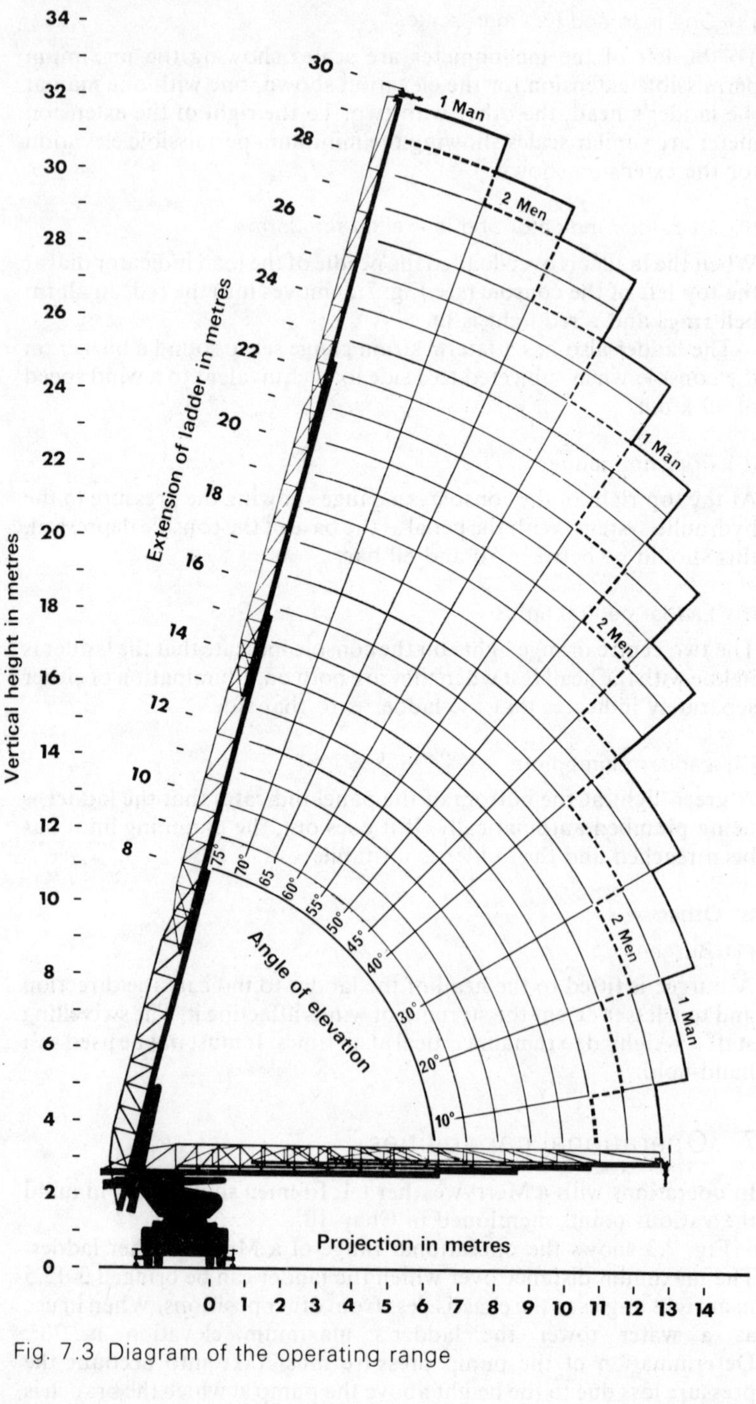

Fig. 7.3 Diagram of the operating range

working. When the ladder is used as a crane the load it can safely lift depends on the angle of elevation, e.g. up to 30°: 450 k.g.; 45°: 675 k.g.: 60°: 900° k.g. Information is given on a panel fixed to the nearside fulcrum frame at the operator's position.

## 8   Hand operating

All ladder movements can be achieved by hand if there is a loss of hydraulic power. (Mechanical damage to the hydraulic pipes feeding the ram locks would however jam them and prevent depression.) The moves are slow and hand operation will therefore usually be carried out only to make up the appliance after a breakdown so that it can be removed for repair.

# Chapter 8
# The Merryweather XRL 30 turntable ladder

The preceding Chapter has described the 'C' type Merryweather t.l. This and similar earlier models have been widely used by British Fire Brigades and many are still operational. Recently however the manufacturers have introduced a completely new ladder, the XRL 30. This differs in a number of ways from ladders of the earlier series and the most important differences are summarised below.

## 1   General appearance and operational capabilities

The table below sets out exactly the main physical characteristics of the XRL 30's configuration and its operational capabilities:

| | |
|---|---|
| Over-all length | 9.7 m |
| Over-all height | 3.2 m |
| Over all width (jacks in) | 2.4 m |
| Over-all width (jacks out) | 3.3 m |
| Maximum extension | 30 m |
| Maximum elevation | 76° |
| Maximum depression | −17° |

Fig. 8.1 shows the ladder's range.

The XRL 30 differs from its predecessors in being a totally self-contained unit with its own engine (see Fig. 8.2). It can be mounted and remounted easily on a wide range of chassis or even mounted as a semi-trailer and towed. The engine is close coupled with a hydraulic pump at the base of the ladder assembly. Around the base is a large circular platform which rotates with it and allows the carriage of personnel other than the operator, for example during the training of a new operator. A cage is being developed for the top of the ladder as are controls to enable operation from the cage.

### a. Jacks

The XRL 30 has two pairs of inclined jacking beams, fitted with self-aligning feet. The controls are on either side of the frame and there is a holding device in the operating valve which enables the operator to leave the console to conduct a visual check. The rear jacks raise the rear wheels clear off the ground after which the front jacks come into

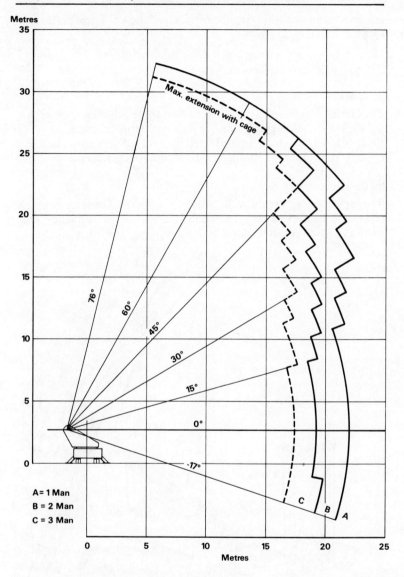

Fig. 8.1 Diagram showing operating range.

action automatically leaving the front wheels down. There are no axle locks.

## 2   Operation of the ladder

### a. Basis

The ladder is powered by an independent three cylinder, four stroke, air-cooled diesel engine close coupled with a hydraulic pump (See Plate 16). All movements are effected hydraulically. A power take-off is being developed as an alternative to the independent engine.

### b. Controls

The jack controls are on either side of the frame. The main controls consist of two levers on a console at the rear of the appliance. One dual axis lever controls elevation and training, the other, dual axis,

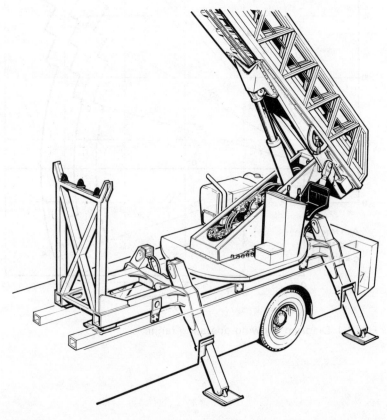

Fig. 8.2  The XRL–30 self-contained unit. The engine on the offside of the turntable provides power to the hydraulic pump.

lever, extension. At the top of these levers are push buttons for the communications system and ladder alignment. Operation to effect any one movement does not affect performance of the others.

### c. Elevation/depression

Both elevation and depression are carried out under hydraulic control. The movements are effected by a dual concentric hydraulic cylinder in which there are two equal rams, one enveloped within the other (see Fig 8.3). In the event of a failure, each can maintain an independent central support of the ladder.

### d. Extension and housing

The extension system features a compact extension winch with ropes of small diameter. The ladders extend and house over oil-impregnated, low-friction sliding pads.

### e. Training

Training is by means of a large diameter single ball race through an epicyclic gear-box and driving pinion. This is connected to a reversible hydraulic motor and integral brake unit.

## 3   Plumbing

The ladder is mounted on a gimbal beam which permits angular movement in two planes. The movement is controlled by two double-acting hydraulic rams connected directly to the beam. These are automatically operated by an electro-hydraulic valve controlled by an inclinometer. The system keeps the ladder plumb vertical on slopes of up to 7° (see Plate 17).

## 4   Safety devices and indicators

### a. Field of operations indicator

On top of the console is a field of operations liquid crystal display unit. This presents precise information relating to load, projection, ladder length, extension height and elevation to safe operating limits.

### b. Locks and cut-outs

Special safety valves stop further movements if the hydraulic system is damaged. There are safety hydraulic locks associated with the training mechanism and cartridge type counter-balanced lock valves to protect the cylinders for elevation/depression against hydraulic failure. Hydraulic safety locks also protect the plumbing rams. The ladder automatically comes to a halt if it encounters obstacles and an alarm sounds if it is over-loaded. In that case all movements towards the cause of over-loading are stopped immediately but the operator can still move the ladder to a safer position.

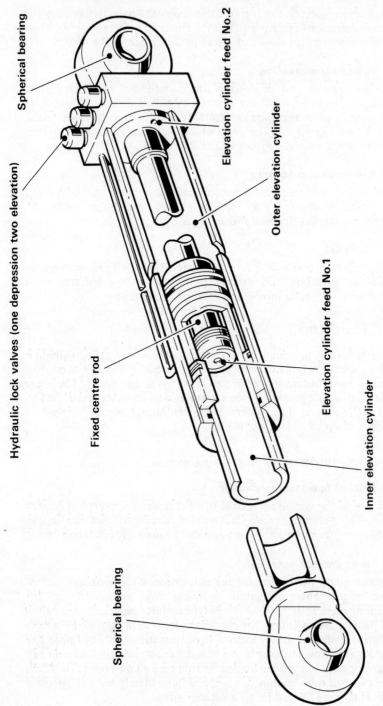

Spherical bearing

Hydraulic lock valves (one depression two elevation)

Elevation cylinder feed No.2

Outer elevation cylinder

Fixed centre rod

Elevation cylinder feed No.1

Inner elevation cylinder

Spherical bearing

Fig. 8.3 Dual concentric hydraulic cylinder for elevation and depression.

# Chapter 9
# The Metz turntable ladder

## 1   General appearance

The Metz turntable ladder consists of a main ladder and three extensions all made from box sectioned steel. The total climbing height is 30 m. The ladder can be elevated to 75° from the chassis and depressed below the horizontal to a maximum of –17° (according to model). There are no pawls, pawling action being achieved by special valves inside the extending rams (see Section 3a. below). Principal controls for the ladder are grouped together on a console mounted on the turntable at which the operator sits.

### a.  Cage

The Metz carries a detachable cage (see Pl.19) which can be removed and carried by two men. This is normally stowed on the underside of the main ladder but for operational use is fitted to the ladder head on brackets and locked into position. At the front is a swing-down, self-locking step-ladder. This allows access from the ground when the ladder is depressed or for rescues at all elevations. At the back of the cage is a gate giving access to the ladder set; this can be locked open or closed. By the use of hydraulic rams the cage is self-levelling at five degree intervals. It has a capacity of 175 kg (two persons) and contains duplicate controls for the ladder movements (see Section 7).

### b.  Other features

Amongst other features of the Metz are:

(i)     a loudspeaker and microphone on the console and top extension;

(ii)    a monitor at the head of the ladder (the monitor can be moved vertically but cannot be trained independently);

(iii)   a fitting for a crane pulley with a maximum lifting capability of four tonnes at the top of the main ladder section;

(iv)    a rotary hydraulic pump driven through a power take-off from the road engine;

(v)     a 225 litre hydraulic oil tank integral with the swinging frame below the heel of the main ladder;

(vi)    jacks with duplicate controls in retractable boxes at the near-side and off-side rear of the appliance. The jacks may be operated separately or together and their movement is infinitely variable. They extend first horizontally outward and then, having reached the limit of their travel, vertically downwards. Full extension of the jacks increases the over-all width of the vehicle from 2.4 m to 4.5 m. When in use they raise the rear wheels completely from the ground.

## 2    Principles of operation

The ladder is operated by hydraulic oil from the rotary pump at a working pressure of 140 bar. Oil passes to control circuits in the fulcrum frame where an automatic control governs the engine speed to give the correct amount of power at any one time. There are four separate circuits for extension/housing, elevation/depression, training, and plumbing. Plumbing is automatic but the other movements are controlled by levers on the console (see Plate 20). Each lever has a constant speed button or c.s.b. Pressure on any of these, after a momentary depression of the accelerator button will bring all control levers into operation at the correct operating pressure. The buttons act as 'dead man's levers' in that release of them stops the movement and returns the lever to neutral. (On some older Metz models there is only one constant speed button separately located on the console). The speed of each operation is controlled by progressive movement of the appropriate lever.

## 3    Operations controlled from the console

### a. Extension/housing

The lever for extension/housing is on the right of the console. The first ladder section is extended and housed by two hydraulic rams, each independent of the other; the other sections are extended by quadruple cables and housed by double cables. Housing does not depend on the force of gravity and is possible at any angle between maximum depression and maximum elevation. The extension lever can also be used to bring the rounds into line manually or the operator can use an automatic aligning switch. Any over-quick housing of the ladder is cushioned by the ram movement which slows down the ladder automatically to prevent damage.

### b. Elevation/depression

Elevation is by twin hydraulic rams. These are locked hydraulically and the ladder cannot be depressed until pressure is supplied to release the lock. The elevation lever is on the left of the console.

#### c. Training

The training lever is at the centre of the console at a lower level than those for extension and elevation. Training is possible through 360° to right and left at all angles above five degrees. It is achieved by a hydraulic motor and chain drive to gearing on the turntable ring.

#### d. Other controls

Other switches on the right of the console control illumination, telephone, round alignment and the transfer of control to the cage (see Section 7, below).

## 4  Plumbing

The ladder is automatically plumbed by a plumb bob which operates two hydraulic rams situated on either side of the pivoted bolster which carries the ladder. Plumbing is effected from 30° of elevation. On more modern models the plumbing gear centralises the ladder in line with the swinging frame when the ladder is depressed onto the headrest.

## 5  Safety devices

Some means of ensuring safety in operations are integral with the various mechanisms of the ladder, e.g. the protection against over-rapid housing (see Section 3a) and the constant speed buttons (see Section 2). There are also the following specially fitted devices:

| | | |
|---|---|---|
| (i) | axle locks and jacks | the axle lock consists of two large hydraulically operated hooks which engage with the rear axle. It operates in sequence with the jacks (see Section 1b). Only when the jacks are fully down and the rear wheels of the appliance off the ground will a safety lock be released to allow the ladder to operate; |
| (ii) | limit stops and locking devices | the ladder stops automatically at the limits of elevation, depression, extension, housing and training. If the ladder has to be operated with one or more jacks in-board it will not be possible to train it beyond 25° from the centre line of the chassis on that side; |
| (iii) | impact and head rest cut-outs | two independent devices are fitted to stop the ladder if it is extended against an obstruction. A device is also fitted to prevent training or depression which |

would bring the ladder incorrectly into contact with the cab or head rest;

(iv)    emergency stop          (see Section 7 below).
        button

# 6    Safety indicators

### a. Loading dials

On the centre of the console is a loading dial (Plate 20) with outer, middle, and inner bands, calibrated in red, amber and green sections. The dial indicates pre-determined limits in relation to jacking positions.

### b. Display panel (field of operation)

Fitted to the underside of the main ladder is a large visual display panel calibrated for elevation and extension (see Plate 8). A moving pointer on this indicates the position of the ladder at any given time. The panel is divided into light green, medium green, dark green and yellow sections to indicate permissible ladder loads.

### c. Inclinometer

A standard inclinometer is fitted to the heel of the main ladder to indicate correct plumbing (see Plate 20).

### d. Lights

Mounted on the ladder set is a panel of four lights: green, amber, red and white. When illuminated the lights have the following meanings:

amber and green    the ladder is approaching its limit and the speed will be slowed automatically;

amber              the free-standing limit has been reached and all controls except housing have been locked. In uses of *extreme urgency only* the free-standing limit can be over-ridden by maintaining constant pressure on the accelerator button. This allows further operation of the ladder (into the yellow section of the display panel) but the ladder can only be loaded if the head is supported;

red                The ladder is at the absolute limit of its use and must not be further loaded (this light is accompanied by an audible warning);

white              the rounds are in line.

A switch on the console can be used to increase the brightness of the lights during day-time operation.

## 7   Remote control

When the cage is fitted to the head of the ladder and the electric circuits connected the ladder operator can press a button on the console to give control of the ladder to the top man. The cage has three operating levers, a load-sensing dial and an emergency stop button. A floor button has to be kept constantly depressed to keep the ladder operative. Maximum speed under remote control is reduced by 50%. The cage has no extension indicator but the entire field of operations is indicated on the load sensing dial with illuminated scales in green, yellow and red.

## 8   Operational range of the Metz turntable ladder

Firemen should bear in mind the general considerations mentioned in Chapter 10 and the points raised above. When the ladder is used as a water tower its maximum elevation is 70°, the two man scale must be used and there must be not more than eight bar pressure at the monitor. In bridging, the ladder can support up to ten people at a time. Fig. 9.1 shows the ladder's general range.

## 9   Hand operating

In case of engine failure all movements can be made by using a manual pump and selecting the circuit necessary (see Plate 21).

| Operation fields | | Jacks fully extended | Jacks in board | Load | Projection | |
|---|---|---|---|---|---|---|
| | | | | | m | m |
| 1 | 1'* | | * | Cage + 2 men | 19.5 | 11.5* |
| 2 | 2'* | | * | Cage + 1 man | 21.5 | 12* |
| 3 | 3'* | | * | 1 man or 100kg | 22.5 | 12.5* |
| 4 | 4'* | | * | When used as a beam maximum ladder extension | 12 men | 16 men* |
| | | | | | Ladder fully extended 26.5 | 15.5* |

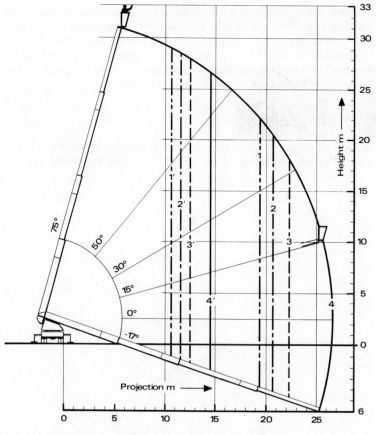

Fig. 9.1 Diagram of operating range.

Plate 1. Metal three-piece short extension ladder.
*Photo: London Fire Brigade*

Plate 2. 10.5m metal extension ladder.
*Photo: London Fire Brigade*

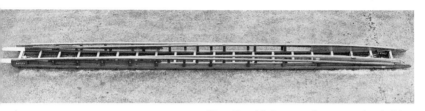

Plate 3. 10.5m wooden extension ladder.
*Photo: Hampshire Fire Brigade*

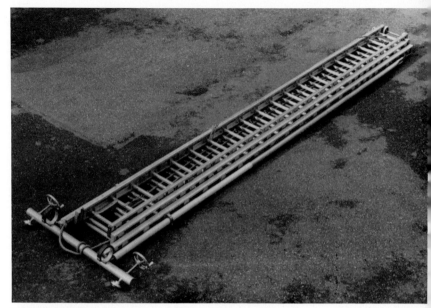

Plate. 4   13.5m metal extension ladder.
*Photo: London Fire Brigade*

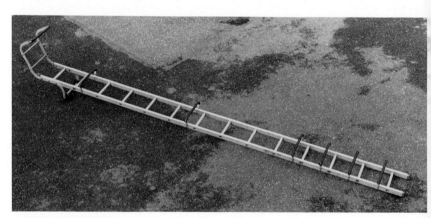

Plate 5.   Roof ladder
*Photo: London Fire Brigade*

Plate 6.   A 13.5m ladder being used to rescue a man trapped in defective window-cleaning cage.
*Photo: Hampshire Fire Brigade*

Plate 7.   A typical detachable two-man cage fitted at the head of a turntable ladder.
*Photo: Avon Fire Brigade*

Plate 8. An illuminated Field of Operations Indicator attached to the main ladder of a turntable ladder.
*Photo: Grampian Fire Brigade*

Plate 9. Magirus DL30 turntable ladder with cage carried on turntable.
*Photo: Oxfordshire Fire Service*

Plate 10. Main controls of Magirus DL30 turntable ladder. Note the field of operations indicator top left.
*Photo: Avon Fire Brigade*

Plate 11. Training a Magirus DL30 turntable ladder by hand.
*Photo: Avon Fire Brigade*

Plate 12. The Magirus DLK 23–12 turntable ladder.
*Photo: Carmichael Fire Ltd*

Plate 13. Magirus DLK 23–12 turntable ladder showing the jacks at maximum extension and the ladder positioned for fitting the cage.
*Photo: Carmichael Fire Ltd*

Plate 14. The "C" type Merryweather turntable ladder.
*Photo: Merryweather and Sons Ltd*

*Plate 15. The Merryweather XRL 30 turntable ladder.*
*Photo: Merryweather and Sons Ltd*

Plate 16. The independent engine fitted to the Merryweather XRL 30 turntable ladder. This is close-coupled with a hydraulic pump to control all movements of the ladder.
*Photo: Merryweather and Sons Ltd.*

Plate 17. The control console of a Merryweather XRL 30 turntable ladder. Note the hydraulic ram on the right, one of two used to plumb the ladder
*Photo: Merryweather and Sons Ltd*

Plate 18.    The Metz DL30 turntable ladder on a Shelvoke chassis.
*Photo: Devon Fire Brigade*

Plate 19.    A Metz DL30 turntable ladder on a Scania chassis with cage
attached. Note the position of the monitor connection.
*Photo: Grampian Fire Brigade*

Plate 20. The control console of a Metz turntable ladder. Note the inclinometer above right and the plate of instructions on using the TL as a crane. *Photo: Devon Fire Brigade*

Plate 21. Using the hand-operated gear to depress a Metz DL30 turntable ladder. *Photo: Devon Fire Brigade*

Plate 22. Metz DL30 turntable ladder. By use of blocks under the jacks the angle of depression has been increased to -25° maximum.
*Photo: Grampian Fire Brigade*

23. Pitching a turntable ladder across the chassis. In areas like this, care must be taken that jacks are not placed on man-hole covers, drain-gratings etc.
*Photo: North Yorkshire Fire Brigade*

Plate 24. A Simon SS 220 hydraulic platform with jacks fully extended. Note the position of the cage in the housed position.
*Photo: London Fire Brigade*

Plate 25. A Simon hydraulic platform illustrating the position of the cage fully housed. Note the fixed piping running along the upper boom and the water collecting head at the rear off-side.
*Photo: Isle of Wight Fire Brigade.*

Plate 26.   Simon SS 220 hydraulic platform cage fitted with folding
platform monitor, hose connections, lighting etc.
*Photo: London Fire Brigade*

Plate 27.   Simon SS 220 hydraulic platform. Control console and boom
controls
*Photo: London Fire Brigade*

Plate 28. Simon SS 220 hydraulic platform showing cage controls and the general construction of the cage.
*Photo: London Fire Brigade*

Plate 29. At low elevations the knuckle can project a considerable distance. Hydraulic platform operators must remember this when training the apparatus.
*Photo: South Glamorgan County Fire Service*

Plate 30. The cage operator's ability to manoeuvre the cage over a building at the limit of the boom is a distinct advantage.
*Photo: London Fire Brigade*

Plate 31. An electrical control panel on a type 'A' emergency tender.
*Photo: Avon Fire Brigade*

Plate 32. Air bags, inflated from a breathing apparatus cylinder, being used to lift a vehicle.
*Photo: Bedfordshire Fire Service*

Plate 33. A typical metropolitan type 'A' emergency tender. The electrical control panel is located on the rear side of the appliance behind the shutters.
*Photo: London Fire Brigade*

Plate 34. A type 'B' emergency tender. This particular vehicle also acts as a BA Tender and a Control Unit. A portable generator can be seen in the rear off-side locker.
*Photo: Cornwall County Fire Brigade*

Plate 35. In areas where there is difficulty in obtaining water, water carriers, such as this example, are invaluable. Carrying 4,500 litres of water and a light portable pump it also has the ability to cross rough terrain.
*Photo: Cornwall County Fire Brigade*

Plate 36.   A breakdown lorry with the facility of lifting and winching 7.5 tonne. Fitted with stabilising jacks, it also carries a range of jacks, air-bags, cutting equipment etc.
*Photo: Durham County Fire Brigade*

Plate 37.   A combined decontamination/chemical incident unit which, following decontamination in an area set out at the back, allows personnel to shower, dress and exit at the front. Airline equipment can be seen at the rear near-side.
*Photo: West Yorkshire Fire Service*

Plate 38.   The decontamination area of Plate 37. Warning signs, spare
cylinders, dam, vacuum cleaner and airlines are visible with clear directions
to personnel.
*Photo: West Yorkshire Fire Service*

Plate 39.   A county foam tender showing the range of gear stowed. 900L
of protein foam compound and 1800L of water are also carried and the
appliance is equipped with two pumps plus HEF generators.
*Photo: County of Clwyd Fire Service*

Plate 40.  A metropolitan county foam tender with a fixed monitor. In addition to 3000L of foam compound, 450L of HEF compound and a Turbex generator are carried.
*Photo: London Fire Brigade*

Plate 41.  A Rolonoff unit showing the pod ready for loading onto the prime mover. This particular pod is a decontamination unit which can be set down and deployed where necessary leaving the prime mover free to be used to move another pod.
*Photo: South Yorkshire County Fire Service*

Plate 42. A demountable pod with electro-hydraulically operated legs which can be extended before the prime mover is removed. This particular pod is a control unit
*Photo: North Yorkshire Fire Brigade*

Plate 43. An adaption of a 4-wheeled caravan to a decontamination unit to be towed into position as necessary.
*Photo: Cleveland County Fire Brigade*

Plate 44.   A divisional control unit used at smaller incidents.
*Photo: London Fire Brigade*

Plate 45.   A comprehensively equipped metropolitan fire brigade control unit. The masts are either retractable or folding and the interior is sub-divided into three compartments [see Plate 46].
*Photo: London Fire Brigade*

Plate 46. The interior of the appliance shown in Plate 45, taken from the communications compartment. Beyond is the working area and, at the far end, the conference section.
*Photo: London Fire Brigade*

Plate 47. A shire county brigade control unit. This is divided into a radio compartment and a working area [see Plate 48].
*Photo: Dorset Fire Brigade*

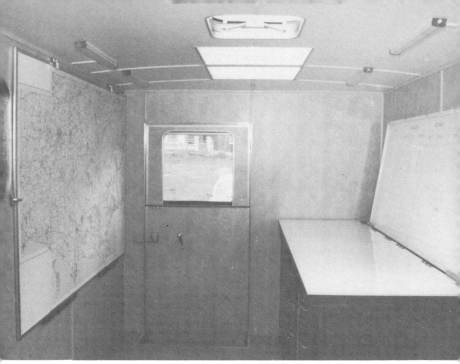

Plate 48.   The interior of the appliance shown in Plate 47. Taken from the radio compartment and showing the county map, mobilising board and perspex working surfaces.
*Photo: Dorset Fire Brigade*

Plate 49.   A typical hose-laying lorry carrying 1,850m of 90mm hose and able to lay single or double lines at speeds of up to 50 kph.
*Photo: West Yorkshire Fire Service*

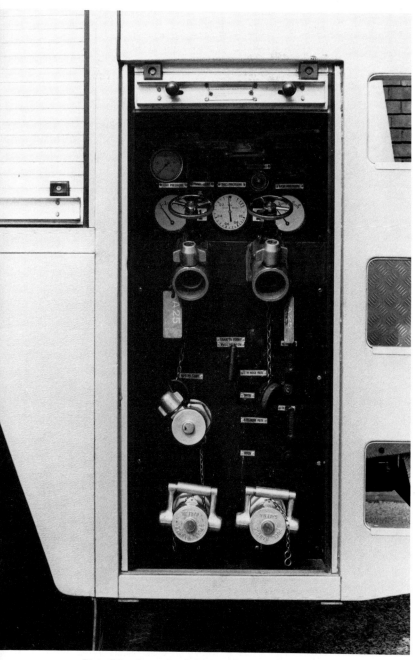

Plate 50. Example of pump side controls to a rear mounted pump, showing gauges, hose-reel valves, high and low pressure controls, tank filling facility and the usual inlets and outlets.

*Photo: London Fire Brigade*

Plate 51. An L4P equipped with a 2250 1/min front-mounted pump. This type of appliance is very suitable for an area with rough terrain and narrow roads.
*Photo: Cornwall Fire Brigade*

Plate 52. A Chelsea side power take off fitted as a "sandwich" type.
*Photo: Shelvoke and Drewry Ltd*

Plate 53. A power take off, "drive-line" type, fitted between the gearbox and axle.
*Photo: HCB-Angus Ltd*

Plate 54. The chassis of an appliance showing the arrangement of water tank, power take off and rear-mounted pump with side controls.
*Photo: Hestair Dennis Ltd*

Plate 55. Brigades vary in their appliance stowage arrangements. Considerable ingenuity is shown in carrying a considerable amount of equipment, readily available, in a comparatively small space.
*Photo: West Yorkshire Fire Service*

Plate 56. Another example of stowage design. Firemen should know the stowage on their appliance but clear labelling obviously helps.
*Photo: Hertfordshire Fire Brigade*

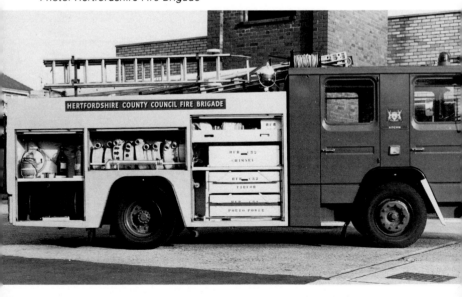

Plate 57. A dual-purpose appliance carrying a 9m ladder, 1365l of water and a rear-mounted pump with side controls. The trunnion bar at the back and quickly removable gantries enables rapid conversion to a pump-escape.
*Photo: London Fire Brigade*

Plate 58. A water-tender ladder carrying 13.5m, 9m, short extension and roof ladders, 1,800l of water and a 4,500l/min rear-mounted pump. Based on a Dennis R133 chassis and driven by a V8-640 162KW diesel engine.
*Photo: Derbyshire Fire Service*

Plate 59. Another example of a water-tender ladder on an HCB-Angus chassis. Carrying 18001 of water, a 13.5m ladder and a 2,2501/,min pump, this appliance is also equipped with a Stemlite telescopic lighting mast.
*Photo: Staffordshire Fire Brigade*

Plate 60. A "compact" appliance based on a Ford "A" series chassis. The tank holds 6751 of water. Pumping capacity 22701/min multi-pressure plus 2 hose-reels, 10.5m, short extension and roof ladders, 3 BA sets and a crew of five men.
*Photo: Cumbria Fire Service*

Plate 61. A Simonitor appliance capable of delivering foam or water at an extension of 12.6m on a turntable. It carries 1,800l of foam concentrate, 135l of HEF and an HEF unit. There are two jacks at the rear to stabilise the appliance as it operates.
*Photo: Essex Fire Brigade*

Plate 62. An experiment to bring the weight of a water-tender ladder below 7.5t. Based on a Bedford KD 120 chassis and powered by a 4900cc diesel, it carries a Godiva 50 multi-pressure pump and 1125l of water. Aluminium is extensively used in the construction.
*Photo: Strathcylde Fire Brigade*

Plate 63.  A pump-escape carrying a 15m steel escape on a Miles
mounting. Water tank capacity 13501 and fitted with a Godiva UMP pump.
*Photo: East Sussex Fire Brigade*

# Chapter 10
# Working with turntable ladders

## 1   Training and responsibility for turntable ladder operation

The ability to operate a turntable ladder successfully must come from highly specialised training and continual practical experience. This Chapter can only outline the principal points to bear in mind when working with t.1s. It may however serve as an introduction to the subject for possible future operators and give an insight into the various factors affecting the use of t.1s. so that they may be employed more effectively on the fireground. The *Fire Service Drill Book* gives some basic practical details on t.1. operations.

Only a fully qualified driver/operator who has successfully completed all the necessary training should be in charge of a t.1. He will then have over-all responsibility and he will man the ground controls. Another crew member may however operate the duplicate controls in the cage. The officer in charge of a fire will decide when he wants a t.1., where he wants it to get to work, and the purpose for which it is required. Usually, however, he will issue only a general direction and leave the operator to interpret it in detail. The operator will normally have greater or at least more recent familiarity with the operational capability and limitations of a t.1. and will be in a better position to judge on the practicability of a particular manoeuvre. The officer in charge should therefore heed his advice.

## 2   Siting a turntable ladder at an incident

When a t.1 arrives at an incident the most important and most difficult task will be deciding where it should get to work. Subsequent elevation and extension will then be relatively easy. The first consideration will obviously be the objective which the ladder has to reach. There may, however, be a number of sites from which the ladder could operate to reach this. In selecting one of them, the operator should bear in mind the following three factors particularly:

(i)     the safety limits of the ladder;

(ii)    clearance around and above the ladder;

(iii)   the state of the ground.

The first objective will of course not necessarily be the only one the ladder has to reach. The ladder should therefore be so positioned as to allow as many other pitches as possible.

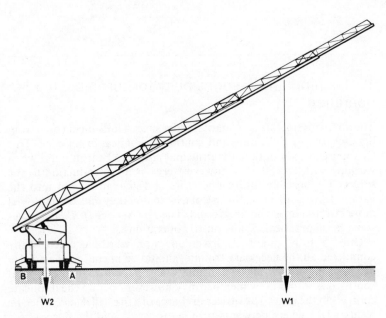

Fig. 10.1 Diagram illustrating factors affecting stability.

### a. Safety limits

(1) Factors affecting stability

Despite the solid working base provided by the jacks and axle lock, where fitted, it is necessary to reduce a ladder's extension when the elevation is decreased. Otherwise there would be a danger that the ladder might overturn. Fig. 10 illustrates this. The weight of the projecting ladder and anyone on it (W1) threatens to overturn the ladder at A. It is counter-balanced by the weight of the chassis, mechanism and ladder to the left of A (W2). The load on the wheel at B (normally referred to as the unburdened near wheel) represents the margin of safety of stability of the appliance and on ladders complying with JCDD 36 should never be less than 1.02 tonnes exactly. Other ladders may have a lower weight limit. This is the basis for indicators given by the inclinometer.

(2) Siting the ladder to ensure stability

The ladder should always be as near the building as possible. In general, this means that the higher the ladder is to be extended, the closer it will be to the building. At heights below 21 m it is possible to operate the ladder safely from further away but a close position is still preferable so that any possibly higher objectives can be reached more easily without moving the ladder.

A procedure for determining whether it is possible to pitch the ladder safely to a particular objective is as follows (see Fig. 10.2). The operator first finds the point B which is the point nearest the building from which he can still see the objective, A. Unless there are other factors to take into account, B then becomes the provisional position for the ladder and the line A–B the ladder's line of pitch. The operator then measures the distance B–C to find the necessary projection (or if this is not possible he measures B–D and estimates the remainder). He then assesses the height C–A and deduces the angle of elevation from the amount of projection. This enables him to ascertain the amount of extension necessary and the inclinometer will indicate whether the particular combination of extension and elevation is within the safe working limits.

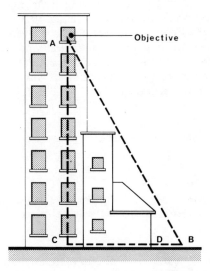

Fig. 10.2 Diagram showing theoretical method of selecting a site for operating a turntable ladder.

A trained and experienced operator should be able of course to decide immediately whether it is possible to pitch a ladder safely to an objective. He must nevertheless still pay close attention to the inclinometer and ensure that he stays within the safe working limits. He must not rely on the safety devices. These are intended only as a fall-back precaution in the event of the limits being inadvertently exceeded. The operator should also bear in mind that in addition to the relation between extension and elevation there is a relation between projection and vertical height increases, so projection decreases. Fig 7.3 show how the various factors which an operator has to take into account inter-act.

In making his judgements the operator will need to assess at least sub-consciously a number of heights, lengths and angles. The height of rooms in a multi-storey block varies but a modern office block of about 30 m would usually have eight floors, a similar block of flats nine. The operator must have practice in assessing the heights of different structures including those without such features as regularly spaced windows that could act as a guide and he must become proficient in pitching to the correct height without hesitation.

### (3) Working on a steep hill

When a turntable ladder is pitched on a steep hill careful positioning is necessary, otherwise the plumbing required will exceed that which the plumbing gear can achieve (see Fig. 10.3 (left)). The operator can overcome this difficulty by approaching at an angle and setting the ladder in a 'semi-staircase' position (see Fig. 10.3 (right)). He should beware of the dangers of over-elevation and over-extension. Close attention to the inclinometer is essential and he should make allowance for the effect on it of the slope, e.g. if the ladder is facing up-hill on an incline of 10° the inclinometer will read 10° though the ladder is still on the head-rest.

### b. Working clearance

Fig. 10.4 illustrates the various clearances needed for a turntable ladder to get to work.

### (1) Jacks

There must be room to lower the jacks. Firemen should bear in mind that they usually extend beyond the wheel-base.

### (2) Ladder heel

There must be room all round the rear of the appliance for the heel of the ladder to travel freely when the ladder is at its maximum projection.

### (3) Overhead clearance

There must be sufficient overhead clearance for the ladder to be elevated from its head rest. Firemen should note that some buildings have projections such as canopies or balconies which render elevation impossible though otherwise clearance is adequate. Elevation is however often possible even where because of the confined space it might not be expected to be. If the obstruction does not project

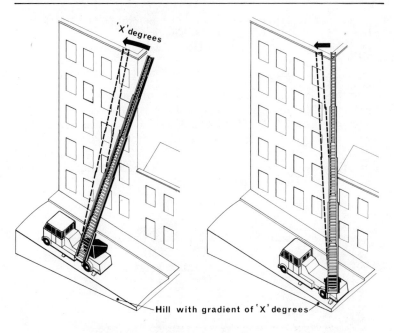

Fig. 10.3 Overcoming the necessity for plumbing in excess of 7 degrees. Left: ladder pitched on a steep hill showing plumbing required. Right: Ladder correctly positioned.

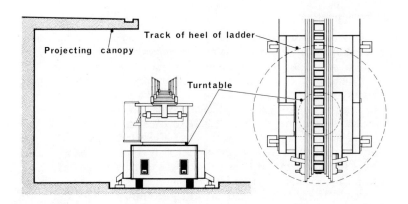

Fig. 10.4 Clearances needed to operate a turntable ladder. Left: Lateral clearance for jacks, vertical clearance for elevation of ladder. Right: Clearance all round the rear of the appliance for the heel of the ladder.

beyond the rim of the turntable it will be possible to elevate the ladder safely to all normal working angles. The operator must bear in mind also that there could be over-head wires which could foul the ladder.

### c. State of the ground

A turntable ladder should only operate on firm ground. Even then there should be an inspection to ensure it is not over hydrant pits, telephone or man-hole covers, or similar plates. This is particularly important in factory forecourts and the drives of private houses where there might be several covers for inspection pits to the various domestic services. Gravel or other surfacing material may often partly cover these (see Plate 23).

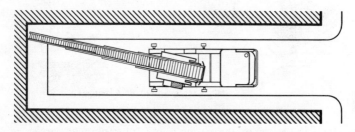

Fig. 10.5 Correct method of getting a turntable ladder to work in a narrow passage.

## 3    General principles of operation

### a. Preliminaries

First, the operator should station the appliance at the position he has selected in accordance with the above criteria. The normal working position is for the ladder to be pitched across the chassis (see Plate 23). This leaves a clear passage for other appliances. The position most favourable to stability however is fore and aft of the chassis. When getting to work in a difficult situation, such as a narrow passage, the operator will do better to back in the appliance and operate the ladder in this position (see Fig. 10.5). Once the t.l. is properly sited the operator should securely apply the hand brake, engage the power take off and axle lock on ladders which have these features then lower the jacks. Where manual jacks are fitted they are lowered by the crew, who also put the chocks into place. Usually, one chock should be in front of one rear wheel, the other behind the other, but on slopes both chocks should be on the down-hill side of the rear wheels. If jacking raises the rear wheels completely from the ground the chocks should be on the front wheels. Jacks must not be adjusted if they lift slightly under normal loading.

## b. Getting to work

An operator should follow this standard sequence of operations in getting a ladder to work:

(i)   he should choose the site for the t.l. by the practical methods described above;

(ii)  he should check the site, including the areas where the jacks will rest, ensuring that the ground is solid and suitable;

(iii) he should drive into position, using other members of the crew to site himself correctly;

(iv)  he should ensure the handbrake is on and, where necessary, engage the power take-off and set the road engine revolutions per minute. (N.B. a ladder, such as the Merryweather XRL 30, has a separate engine to power the ladder);

(v)   he should carry out axle locking, jacking and chocking in the manner laid down for that particular appliance, according to the particular circumstances;

(vi)  he should carry out the necessary ladder manoeuvres for the required pitch, observing safety procedures. This will involve: elevating the ladder to clear the gallows and attain the approximate angle required; training to the necessary point; checking the plumbing is correct; reading off the permissible extension for that elevation from the inclinometer or field of operations indicator; extending to the permitted limit and either resting on the pawls or seeing that rounds are aligned; adjusting the ladder finely and slowly to achieve the exact position required.

N.B. On each occasion, before extending or housing a t.l., the operator should visually check that there is nobody standing on the ladder other than on the platform, showing a clean pair of heels, or in the cage.

## c. Making up

The operator should likewise follow a definite sequence when making up:

(i)   he should bring the head of the ladder clear of the building. Usually he will do this by training but elevation or housing may be necessary, for example if the head has been inserted into a narrow opening;

(ii)  he should house the ladder;

(iii) he should train the ladder to a central position over the appliance;

(iv)  he should depress the ladder to within 10°–15° of the gallows;

(v)    he should check the plumbing, if necessary;

(vi)    he should depress the ladder onto the gallows;

(vii)    he should house the jacks and remove chocks and axle locks.

# 4   The use of turntable ladders

### a. Functions of the turntable ladder

A turntable ladder is used most commonly as a water tower or to effect rescues from tall buildings but is available for any purpose which requires an aerial platform e.g. as an observation point. It can also be used to provide an external staircase or in bridging.

### b. General considerations

#### (1) Operating with a man aloft

It will in many cases be necessary to operate the ladder with a man in the cage or at its head. He can either climb up when the ladder is extended or mount first and be taken up with the extension (this is known as 'shooting up').

The operator should not operate the ladder until the man aloft has signalled that he is ready by extending his left arm horizontally to its full extent. He should make all adjustments at slow speed. A man climbing a turntable ladder at normal elevation should do so as he would any other ladder, that is, with his hands on the rounds. At low elevations, however, it is more convenient to use the hand rails. To avoid undue oscillation firemen should not climb the ladder in a uniform rhythm. If it is necessary to pass an object up the ladder the man higher up should descend to the one below to collect it.

#### (2) Making adjustments

Adjustments to the position of a t.l. may affect the safety limits. The operator must accordingly watch the inclinometer and check the plumbing. He must be alert for audible and visual warnings from the safety indicators. He must ensure that the ladder pawls, where fitted, are brought to rest after adjusting the ladder's extension. If it is necessary to adjust the ladder's extension with a man at its head, the operator should ensure the man is safely on the platform, hooked on, with toes clear, and aware of the intended adjustment.

#### (3) Support

When the ladder is used with its head resting against an objective ('rest down'), the operator must ensure that the objective is likely to be sufficiently solid, that the ladder is only lightly touching, and that it rests squarely. To take up this position the operator should elevate the ladder to a somewhat steeper angle than necessary, then extend it. When he has adjusted the training as necessary he should ease the head of the ladder gently down, taking care that both strings of the ladder touch at the same time to avoid distortion. In this position a load of eight people is usually permissible provided they are evenly distributed over the whole ladder length. In bridging, the head of the ladder should be rested if possible. This will allow a greater load to be supported provided it is evenly distributed along the ladder.

### (4) Movement

An operator must not move his appliance while the ladder is extended or there are men on it.

### (5) Guy lines

Guy lines should always be used on a ladder in high or gusty winds, two men attending each. Firemen should appreciate that even in generally calm weather there may be strong local winds, due for example, to the presence of tall buildings; the wind may also be stronger at the head of a ladder than at ground level. The lines should never be made fast and should be held at as wide an angle as possible in order to reduce vertical pull on the ladder. The strain on the guy line should only ever be enough to counter the effect of the wind. Where possible the operator should site the ladder so that the wind does not hit it from the side.

### c. The turntable ladder as a water tower

All turntable ladders have a monitor at their head or the front of the cage (see Fig. 4.2 and Chapter 4, Section 5a). The monitor can sometimes be controlled by ground lines but is usually operated from the head of the ladder or its cage. Ground lines should always be used in strong winds or other dangerous conditions and no-one allowed on the ladder. They should also be used when the elevation of the ladder needs to exceed 70° (on ladders where this is permissible). The lines are for vertical control of the monitor only: horizontal adjustment is by training.

Firemen should bear in mind the follow points:

(i)     they should only use a turntable ladder as a water tower within the limits indicated on the appliance. The two man scale should be used;

(ii)    the ladder should be so positioned that falling debris is not a danger;

(iii)   the best distance for a monitor is 3.5 to 4.5 m from the objective. Further away would make the jet less effective, closer would bring the man at the head too close to the fire;

(iv)    a pump supplying a t.l. monitor should not be supplying any other branch and should preferably have a completely independent water supply. No other jets should be at work from a t.l.;

(v)    there must be a means of reducing the working pressure of the monitor jet. This is usually a special control valve inserted in the hose line. The pressure must not be released too suddenly, however, as the loss of pressure would inevitably cause the ladder to sway considerably. Hose must not be allowed to hang unsupported. The maximum pump pressure for nozzles from 19 mm to 41 mm is 9.5 bar and for some t.1s. the manufacturers suggest a lower maximum pressure for the larger sizes;

(vi)   only in exceptional circumstances should firemen use a t.1. as a water tower near high voltage electricity cables. The absolute minimum distances any part of a t.1. should be from cable are: seven metres from cable carrying current at 400,000 v.; four metres from cable at 32,000 v.; and three metres from cable at 25,000 v. The ionising effect of smoke and the conductivity of a water jet can greatly reduce these distances.

### d. The turntable ladder in rescues

In almost all turntable ladder rescues those rescued are capable of being assisted to walk down the ladder. When the ladder is being used to rescue such people from a window or balcony the operator should site it in a staircase position with the strings as close and as parallel to the face of the building as possible (see Fig. 10.6). Those to be rescued can then step sideways onto the ladder which will be easier and more

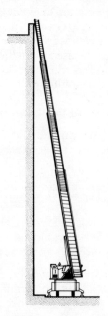

Fig. 10.6 Placing a turntable ladder to effect a rescue.

likely to inspire confidence than having to manoeuvre themselves around it. Those rescued should always be told to grasp the ladder rounds and a fireman should precede them down, giving whatever guidance and support is necessary. Firemen should not attempt to carry down on a t.l. as the trussing renders the operation unsafe at most angles of elevation.

When those to be rescued are not capable of climbing down the ladder, the rescue cage can be used. Otherwise firemen should lower them down in a sling, making use of the rescue equipment that is fitted to the head of the ladder (see Fig. 4.3). In so doing they should follow the procedure laid down in Drill T4 of the *Fire Service Drill Book*. The operator must not extend or house the ladder once a person is in the sling. He should swing the person clear of the building by training the ladder away. The tail line and guy line attached to the ring of the rescue sling can be used to steady the person during the rescue.

Because of the loads and stresses imposed on the ladder during such a rescue the limits for two man loading should apply and the operator must not exceed these.

## 5   Service and maintenance

Routine servicing of t.ls. — cleaning, oiling, greasing etc. — should take place at stations in accordance with normal Brigade procedure. Maintenance however is for qualified staff only.

Standard tests, as prescribed by the *Fire Service Drill Book,* should be carried out on acceptance, quarterly, after operational use, and whenever else it is considered necessary. Qualified personnel should also carry out periodic technical inspections, preferably every six months. These should check the condition and behaviour of all essential parts of the ladder, its mechanism and safety devices. They may reveal the development of a fault at an early stage when it can easily be rectified. The use of report forms in these inspections ensures that no important items are over-looked and provides a useful record for future reference.

# Part 4
# Hydraulic platforms

## Introduction

The adaptation of hydraulic platforms from commercial to Fire Service use took place in 1961, since when there has been continuous development. Most hydraulic platforms used in Britain are British made but a few are continental. They are mounted on various chassis. Any pump they carry will be either portable or completely independent of the road engine. Earlier types had two booms and measured between 13 m and 15 m to the base of their cage at full height extension; since then, however, heights and projections have steadily increased to produce the modern 26.3 m and 30 m types.

Hydraulic platforms complement rather than replace turntable ladders (including the modern type with a cage). Each has certain advantages and certain disadvantages. A hydraulic platform provides a very stable base for various Brigade operations: rescue, large monitor attack, b.a. crew entry, ventilation etc. The ability of the three boom type to reach over a building to pick up or set down and to move up to five persons at a time is also useful.

This Part describes the general characteristics of hydraulic platforms and principles of working with them and it looks in detail at some of the more common types.

# Chapter 11
# General information on hydraulic platforms

## 1   Terminology

As with turntable ladders, firemen have adopted a standard terminology for use when operating hydraulic platforms.

| | |
|---|---|
| Booms | the two or three jointed sections which carry the cage. |
| Cage (or platform) | the personnel compartment at the end of the second or, if fitted, third boom. |
| Depress | to reduce the height of the cage. |
| Elevate | to increase the height of the cage. |
| Height | the distance of the cage bottom from the ground. |
| Knuckle | the pivoting joint between booms. |
| Over-ride | a control at the base operator's position. |
| Plumbing | use of the jacks to compensate for any camber up to five degrees — 1 in 12 — and bring the vehicle level. |
| Projection | the distance from the outside of the jack foot to the outside edge of the cage when the bottom boom is fully elevated and the second boom horizontal, at right angles across the chassis. |
| Safe working load (s.w.1) | the specific payload which an h.p. can normally carry anywhere within its working range. It can be affected by the vehicle being unlevel, strong winds, or the imposition of extra loads on the booms, e.g. by use of the monitor. |
| Train | to move the cage in a circular route by moving the turntable. |
| Turntable | the revolving platform, exactly on the centre line of the chassis, which carries the fulcrum frame and one end of the bottom boom. |

## 2 Design of hydraulic platforms

### a. Specifications

Two specifications apply to hydraulic platforms. JCDD 27, published 1971, revised November 1975 and amended December 1980 refers to appliances between 19.8 m and 22.9 m; JCDD 34, published November 1975 and amended December 1980, to appliances between 24.4 m and 30.5 m. These specifications lay down certain maxima and minima only. Manufacturers may choose to exceed them or Brigades to set additional or stricter requirements, perhaps to meet special circumstances. Like turntable ladders, hydraulic platforms must also comply with current *Motor Vehicle (Construction and Use) Regulations* and the *Road Vehicle (Lighting) Regulations.* This Chapter describes only the general characteristics of appliances complying with the JCDD specifications. Any fireman wishing to make a more detailed study should consult the specifications themselves.

### b. General description

Hydraulic platforms consist essentially of two or three booms hinged together. The two lower booms pivot in a vertical plane on each other and on the fulcrum frame, on which the bottom boom is hinged. The third boom takes the form of a pivoted or telescopic extension arm at the upper end of the second. The fulcrum frame is mounted on a turntable on the centre line of the chassis, over the rear axle. At the head of the top boom is a hinged platform, the floor of which gravity, or a hydraulic ram, keep constantly parallel to the chassis. When the platform is fully housed the knuckle between upper and lower boom rests on the driver's cab, whilst the cage and third boom, if fitted, are at the rear of the appliance (see Plates 24 and 25). Hydraulic motors driven by a power take-off from the road engine effect all movements, including that of the jacks. The hydraulic oil is usually stored in a tank on the sub-frame. Overall length, width and height, turning circle, acceleration, braking, road holding etc. vary according to type. The relevant specifications set out the requirements to be met (See Fig 11.1).

### c. Controls: booms and turntable

The main control console is on the turntable to one side of the bottom boom. There are duplicate, though modified, controls in the cage. The main console operator can however override these in an emergency.

### d. Jacks and plumbing

Individual controls at the rear of the appliance operate each of the four jacks. The jacks are moved by hydraulic rams and, on the longer appliances, can lift the rear wheels off the ground to provide a stable

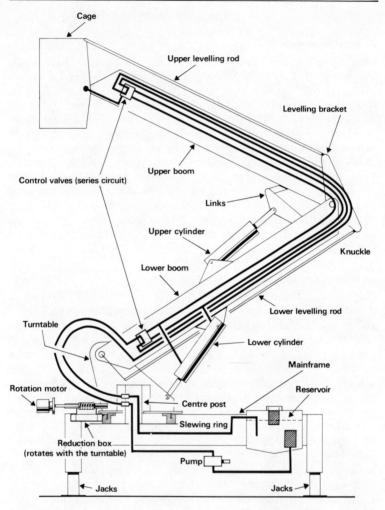

Fig. 11.1 General diagrammatic sketch of parts of a hydraulic platform.

platform. Axle locks are unnecessary as is plumbing gear. Adjustment of the jacks effects any necessary plumbing, to a maximum of five degrees.

### e. Cages

These vary, but the largest can accommodate five adults. Access is by hinged steps from the ground and from the front of the cage. A cage consists basically of a metal platform, fitted with controls for the booms and turntable, railed round for safety, and hinged to the top boom. Some types also have a road engine starter button. There is pipe-work to a monitor and one or two female outlets for firefighting; underneath, there is often also a water spray attachment to protect personnel in the cage from heat. Some cages have a front fold-down section about 600 mm wide which can carry either persons or equipment such as a stretcher. Brigades also sometimes have fitted such extra equipment as hydraulic connections for power tools, b. a cylinder supplies, or lighting.

### f. Safety features

These include means:

(i)     to prevent movement of the booms or turntable until the jacks are firmly on the ground;

(ii)    to prevent retraction of the jacks until the booms are stowed;

(iii)   to prevent unsafe positioning of the booms;

(iv)    to hold the booms in place if the hydraulic system or engine driving the hydraulic pump fails;

(v)     to control the elevation of the booms in relation to one another and see that their limits are approached slowly;

(vi)    to provide an auxiliary motor for the booms, turntable and jacks if the main road engine fails;

(vii)   to allow the base operator to override the cage controls.

### g. Communications

All hydraulic platforms have provision for the base and cage operators to communicate.

### h. Pumps

Any pumps on an h.p. must be self-contained, with an independent engine. This allows individual control of pump and platform and generally facilitates removal of the pump for maintenance.

## 3    Acceptance tests

The JCDD specifications lay down tests, which can take place anywhere agreed between the manufacturer and Brigade. If, however, the manufacturer can provide a certificate guaranteeing that another appliance of the same design and construction has already passed the tests, they can be omitted.

# Chapter 12
# The Simon hydraulic platform

## 1   General appearance

There are three modern types of Simon hydraulic platform: the
SS220, the SS263 and the SS300. All are of a similar three boom
design and have the following maximum working heights: SS220,
23.5 m; and SS263, 27.8 m; SS300, 31.5 m. (The SS70 and SS85 are
older models sometimes still found in use. They have maximum
working heights of 21.3 m and 26 m respectively but otherwise are
generally similar to the new range). Fig 12.1 shows the general
conformation of the appliances. The mainframe can be mounted on
various chassis, according to Brigade requirements. It is of welded
steel, rigid and bolted to the chassis frame, incorporating the
stabilising system. At its rear end is a heavy steel turntable which runs
on a double row ball-bearing slewing ring on steel tracks. This can
rotate 360° in either direction. On it are pivoted the lower end of the
bottom boom, on which are located the boom controls, and the
hydraulic cylinder (or, on the SS300, twin hydraulic cylinders) which
operate the boom. Other main controls (except those for the jacks),
gauges, etc. are mounted on a console to the left of the boom controls.
Here the operator sits and is carried round by the turntable so that he
can observe all movements.

### a. Booms

The booms are of a steel box construction to give rigidity and lessen
any tendency to whip. All necessary piping and wiring is carried
externally in channels to minimise heat or mechanical damage. The
first, hermetically sealed, boom can move through 84°. On its upper
end is pivoted the second boom, which can move through 150°. The
hydraulic cylinder which operates it connects the two (the SS300 has
two cylinders). The third boom can be raised to align with the second
or lowered by gravity until fully vertical. For travelling, an automatic
valve makes it possible to power this boom 30° past vertical to
minimise the length. Over-all lengths of the various Simon models
are: SS220, 9.3 m; SS263, 11.6; and SS300, 13.5 m.

### b. Cage

The cage frame is made of a high strength light alloy and has a floor
area of 1.85 m². The floor, of an expanded alloy, is designed to carry

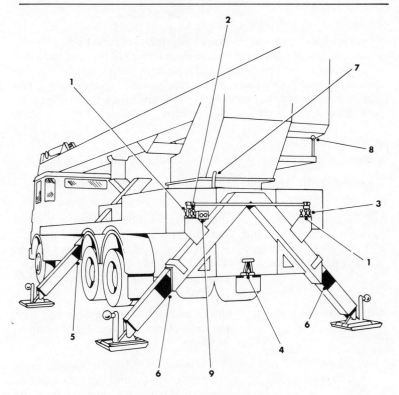

1   Selector valve
2   Control valve for left hand jacks
3   Control valve for right hand jacks
4   Level indicator
5   Front jack spread indicating diamonds
6   Rear jack spread indicating diamonds
7   Turntable stowage position indicator
8   Power system "energised" indicator light
9   Filter indicator

Fig. 12.1  General sketch of Simon hydraulic platform.

equipment and up to five adults as far as the specified load limit. The cage has a 1.1 m high handrail. Regular access to it is from a two-step ladder on the near side or from a fold-down platform at the front. This platform also has a folding handrail; it is about 600 mm wide and gives an extra area of just under one square metre. A stretcher can be fitted to the cage. The cage has its own controls for operation of the booms and the turntable and is kept permanently level automatically (see Plate 26).

#### c. Jacks

There are four telescopic, hydraulically operated jacks to stabilise the appliance, one at each corner of the main frame. They are constructed of a rectangle steel box section and have large swivelling end plates to spread the load evenly and raise the rear wheels from the ground. The jack controls are at the rear corners of the appliance. Extension of the jacks doubles the appliance width from 2.45 m to 4.9 m.

#### d. Other features

Among other features of the Simon hydraulic platform are:

| | | |
|---|---|---|
| (i) | a 400 litre hydraulic oil reservoir | this is located in the mainframe; |
| (ii) | a manually controlled monitor and, to give a water spray curtain, a fixed nozzle | the monitor is mounted on the front of the cage, the nozzle beneath. Further information is given in Section 4 below. There is also the option of a remotely controlled monitor as an alternative to the manual; |
| (iii) | a shackle type lifting eye | this is located under the cage and with the cage empty will take loads of up to 365 kg; |
| (iv) | means of communication | this consists of a handset on the main control console and a microphone/loudspeaker in the cage so that the operator can communicate with the man at the top; |
| (v) | a stand-by operating system | see Section 2a. below; |
| (vi) | light alloy ladders | these optional extras fold up to the sides of the booms and incorporate a self-levelling cross-over platform at the lower knuckle; |
| (vii) | various other optional extras | examples, in the cage, are: connections for hydraulically powered tools, compressed air lines, and electric power connections. |

## 2   Principles of operation

The Simon is hydraulically operated by oil from the 400 litre oil reservoir. Engagement for the pump which drives this varies from vehicle to vehicle. Rotation is by a hydraulic motor through a reduction gear unit; movement of the booms is by hydraulic cylinders. These decelerate automatically at the end of each stroke thus bringing the booms to a smooth and gentle halt. The circuits are so designed that there is no loss of speed when more than one boom is in operation. Speed is controlled by the position of the appropriate control lever. Movement is not possible until the jack controls have been operated. Resistance to the jacks caused by firm contact with the ground causes the pressure in the jack cylinders to rise and hydraulic valves close thus allowing an oil supply for the boom movements.

### a. Emergency operation

If the main engine or hydraulic pump fails the operator can call upon a stand-by nine kilowatt petrol engine which drives an auxiliary hydraulic pump. This renders all operations possible at a reduced speed from either the turntable or cage controls. In addition it is possible to rotate the turntable by hand cranking through an extension shaft on the worm gear and the booms can lower by gravity if the cylinder lock valves are opened manually. The jacks can be levered up by hand if the weight is lifted from them and the oil emptied from their cylinders.

## 3   Controls

### a. Main controls

On the turntable mounted console there are a rotation control lever, a cage control shut-off lever (by which the operator can immediately override the cage controls) and controls for communications with the cage. To the right, on the lower end of the bottom boom, there are three levers to control the booms (see Plate 27).

### b. Cage controls

The cage operator's controls (see Plate 28) consist mainly of two joystick levers which control all boom and turntable movements. Operating instructions are laid out pictorially. Other controls are switches for spotlights, push-buttons for the remote control monitor if fitted, a control for the intercom system, and, importantly, a lever which effectively neutralises all cage controls while the cage operator is carrying out rescues. This eliminates the possibility of accidental operation.

### c. Jacks

The jacks are operated by handles at the two rear corners of the vehicle. The jack tubes are marked with black diamonds and at least half of each diamond should be visible to ensure that there is sufficient jack spread for stability. (Fig. 12.1)

## 4   Water system

The appliance can carry a portable pump of 2275.1/min capacity at 7 bar. This or a supply pump can feed a standard collecting head under the rear bodywork. An inlet relief valve limits the maximum pressure to ten bar. By a system of light alloy tubing, swivel joints and heavy duty hose at the knuckle joints the water is fed up to a monitor at the foot of the cage. This usually has a jet/fog nozzle and is capable of a 90° sweep either side of centre and 45° elevation and depression. The full flow of water can be taken at any angle of the monitor and any position of the booms. A manifold on the monitor mounting incorporates an additional 63.5 mm outlet on one side and a hose reel connection on the other. A lever valve at the monitor controls the water curtain nozzle under the cage.

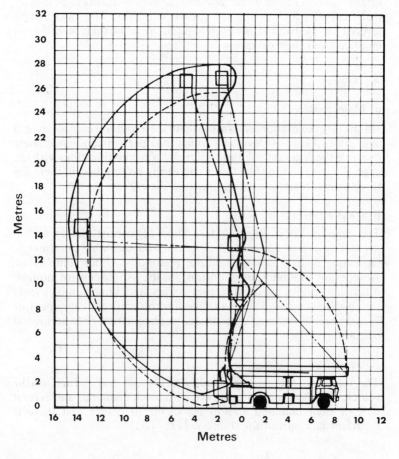

Fig. 12.2(1) S.S. 263.

94

## 5   Safety

Certain safety features are separately operated: the control neutralising lever in the cage for example has already been mentioned. The jacks cannot retract other than under hydraulic power (unless manually operated) and then only when the booms are stowed. Automatic limit stops make it impossible to exceed safe working limits. Visible or audible warning devices are therefore unnecessary. There are however pressure gauges on the control console and a hydraulic pressure warning light at the cage controls.

## 6   Operational capabilities

In working with a Simon hydraulic platform firemen should bear in mind the general points made in Chapter 14. Fig 12.2 shows the

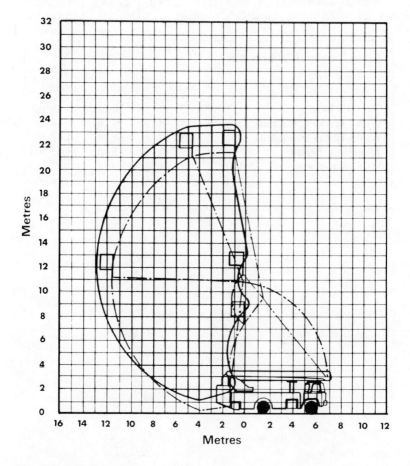

Fig. 12.2(2) S.S. 220.

operational range of two models, while the tables below summarise the exact working ranges and over-all dimensions of the three modern types.

| Table 1: working ranges | | | |
|---|---|---|---|
| | SS220 | SS263 | SS300 |
| Maximum working height | 23.5 m. | 27.8 m. | 31.5 m. |
| Maximum height to cage floor | 22.0 m. | 26.3 m. | 30.0 m. |
| Maximum outreach | 12.5 m. | 14.5 m. | 15.3 m. |
| Cage safe-working load on hard level ground with pipes dry | 365 kg. | 365 kg. | 365 kg. |
| Cage safe-working load with monitor delivering 2275 1/min. with 35 mm nozzle | 227 kg. | 227 kg. | 227 kg. |

| Table 2: over-all dimensions | | | |
|---|---|---|---|
| | SS220 | SS263 | SS300 |
| Over-all length | 9.33 m. | 11.58 m. | 13.48 m. |
| Width (jacks retracted) | 2.44 m. | 2.44 m. | 2.44 m. |
| Width (jacks extended) | 4.88 m. | 4.88 m. | 4.88 m. |
| Travelling height | 3.25 m. | 3.43 m. | 3.43 m. |
| Minimum g.v.w. | 14.73 t. | 15.23 t. | 16.26 t. |

# Chapter 13
# Working with hydraulic platforms

## 1 Training and responsibility

Operating hydraulic platforms is a highly skilled job which can only be undertaken after specialist training. Even then, continuing practical experience of all types, in all circumstances, is essential for the hydraulic platform operator to develop his skills. This Chapter can only outline some general points which firemen should bear in mind; the *Fire Service Drill Book* gives further practical details.

A hydraulic platform at work should always have two qualified operators: the operator in the cage and the driver-operator at the ground controls. Although the cage operator will usually carry out most of the manoeuvres once the appliance is in position, it is the driver who will have the over-riding responsibility for the platform and its safety. The officer in charge of the incident will decide when and where he wants a hydraulic platform and what he wants it to do, e.g. effect a rescue or entry, get a monitor to work. The driver however will decide if the proposed manoeuvre is practical. If it is, he must execute it according to his own judgement. He must maintain communication with the cage at all times, watch all its manoeuvres and be ready to take over in an emergency.

## 2 General considerations

Having positioned his appliance in the safest and most convenient location for the operation in hand, the driver-operator should ensure the parking brake is firmly on, engage the power take-off and set the hand throttle. The chocks must be in place — one in front of and one behind the front wheels on level ground, but both on the downhill side on a slope — and they must not be adjusted once the booms are in operation. A long line should be taken aloft in the cage to haul up extra gear as necessary. At the end of operations, it is important especially in freezing weather to drain the water pipes.

### a. Working range

It is important for the operator to be able to judge, reasonably accurately, at what distance from a building he needs to place his hydraulic platform so that the cage can reach all the points to which access may be necessary. He should memorize the working range for his particular appliance (see Fig 12.2). Generally a hydraulic platform reaches its maximum height at a radius which puts the cage just outside the jack feet with the booms across the vehicle. Maximum radial projection occurs at approximately half maximum height. Operators must during their manoeuvres also take care that the knuckle boom can travel safely. At low elevations particularly it can project a considerable distance (see Plate 29). They may need, in a narrow road, to raise the cage first to its maximum height before rotating it 180°. They must also bear in mind the spread of the jacks.

### b. Ground conditions

Obviously the ground must be capable of supporting the appliance and not allow the jack feet to sink. The maximum jacking loads occur, not during jacking up, but as the booms are rotated at full radial projection. Ground which will support initial jacking loads may therefore not support the load during operations. The operator should watch for:

(i)     drains or man-hole covers;

(ii)    edges of embankments, canal banks, etc;

(iii)   edges of roads, kerbed or not, where there is no support beyond the edge.

Generally, ground which will support the appliance without the wheels causing indentations will be satisfactory but the operator must make allowances for multiple axles, special tyres etc. He must also remember that surface conditions can change during operations. Rain or water from fire fighting can soften the ground; frozen ground can thaw; and, on a hot day, a bitumen surfaced road can melt.

## 3   Using a hydraulic platform on unlevel ground

Hydraulic platforms do not have plumbing gear. Accordingly, when the ground has a camber (slope across the appliance) or a gradient (slope along the appliance), the operator must use the jacks to bring the vehicle level. Jacks must be extended before the booms can be operated.

### a. Cambers

Levelling is usually possible on cambers of up to five degrees, i.e. 1 in 12. Where the design is such however that the rear wheels are lifted from the ground, the vehicle suspension may limit this. If levelling is not possible, even with correct 'packing' under the jacks, the operator should position his appliance to work on the 'up-hill' side.

### b. Gradients

On level surfaces an appliance will usually be jacked with its rear end high, since the rear wheels are normally lifted clear of the ground. If the operator positions his appliance facing up-hill he can use this characteristics to bring it level or at least lessen the effect of the slope. It is always better to operate the cage up-hill since, if it is operated down-hill from the turntable, it may be impossible then to rotate it up-hill when loaded, without elevating it to its maximum height.

## 4    Safe working loads (S.W.L.)

A hydraulic platform can normally carry its specified s.w.l. throughout its whole range of movements. A reduction of the s.w.l. is, however necessary when the appliance is:

(i)    not levelled;

(ii)    used in strong or gusty winds (see Section 5b below);

(iii)    subjected to extra loads on the booms.

Use of the monitor can impose an extra load. Each type of h.p. has restrictions laid down, depending on the amount of water being delivered and the amount held in the boom pipes. A relief valve is usually fitted to the water system to ensure that, if the monitor is used with the recommended nozzle, the maximum permissible reaction is not exceeded.

Other extra loadings occur when the retractable front of a cage is used to carry a load or, when ladders are fitted to the booms, they are used for any purpose. The ladders themselves also have loading restrictions.

## 5    Hazards

### a. Physical hazards

As already noted, the operator must ensure that his appliance is on firm ground and must make allowance for the room needed to manoeuvre. He must look out for power lines, telephone cables, etc., especially at night: use of the searchlight can help. Live conductor wires are a particular hazard and firemen may find the Health and Safety Executive's *Guidance Note GS6* 'Avoidance of danger from overhead electric lines' valuable reading.

In addition to these possible dangers, crews will often find various haphazardly sited and potentially dangerous obstructions when they are placed onto a roof or level area on top of, or over, a building. The cage operator has a distinct advantage here in 'driving' at the limit of the booms (see Plate 30).

### b. Wind conditions

Among the general conditions operators must assess to ensure safe working of appliances are wind forces. The main factors are boom position, pay-load, speed and gustiness of the wind, and the general working environment. Operators must take into account the peculiar effects produced by conglomerations of tall buildings. They must also compensate by lighter loading for any increased wind force on the cage resulting from a special design, such as closed sides or a canopy. The Beaufort scale is a general guide to wind speeds and effects. Firemen must however remember the special dangers of gusty conditions: a sudden gust can be more hazardous than a strong but steady wind. Fig 13.1 suggests, at least for Simon hydraulic platforms, estimated safe loadings for various wind speeds and size of h.ps. Operators may increase these in sheltered conditions, provided they bear the other factors mentioned in mind. They should remember too that in operation the cage may be elevated out of, or above, earlier shelter and be subjected to a wind force.

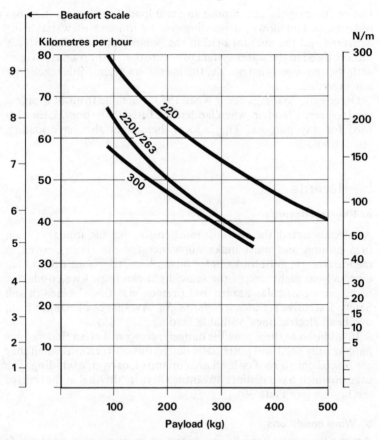

Fig. 13.1 Example of estimated safe loadings for various wind speeds on Simon hydraulic platforms.

## 6   Maintenance

Routine servicing, such as cleaning and lubrication of the boom mechanisms in accordance with the manufacturers instructions, should take place at station level. Further maintenance is for qualified staff only. H.ps should be regularly tested in accordance with the *Fire Service Drill Book.*

# Part 5
# Special appliances

## Introduction

To the Fire Service a special appliance is any vehicle other than a normal pumping appliance. The term includes the turntable ladders and hydraulic platforms dealt with in Parts 3 and 4 of this Book.

Originally pumping appliances carried all the equipment firemen used and this was generally adequate for dealing with most ordinary fires. However there were always fires that presented greater problems such as those where there was a shortage of water or where the materials involved posed some special risk. These fires increased in number as new industrial processes were introduced, new materials came into use and new developments took place in a wide range of areas across society at large. Fire Brigades also began to attend more special service calls, particularly to effect rescues. The extensive developments in chemical transport are one special source of incidents, involving problems of handling special risks, of control and of decontamination.

New working methods and new items of equipment were developed to meet these new situations. To carry all this equipment on all pumps was not practical and so various special appliances were devised. Their number is considerable and the particular design of each type varies as they depend very much on the individual preferences and problems of different Brigades. JCDD specifications have been prepared for emergency tenders only.

This part looks at the general features of some of the more usual special appliances in their basic forms. These forms may be adapted and firemen should note that, for convenience, special appliances are often designed to serve a dual function. The combination of a foam or emergency tender with a salvage tender is particularly common. Some other combinations firemen may encounter are: emergency tender/control unit; emergency tender/chemical incident unit; foam tender/water carrier; control unit/chemical incident unit; control unit/canteen van. Some Brigades are also experimenting with the use of demountable pods, trailers and other means of enabling special equipment to be brought to an incident without commiting an appliance permanently and exclusively to that role. These are also looked at briefly.

# Chapter 14
# Special appliances

## 1    Emergency tenders

An emergency tender, to the fireman, is an appliance specifically designed to carry a wide range of special equipment for use both in firefighting and on special service calls.

### a. Specifications

The Joint Commitee on Design and Development has prepared specifications for two types of emergency tender Type A, JCDD 9 and Type B, JCDD 8. Both specifications were last amended in 1977. Like other JCDD specifications these lay down minima and maxima only and allow scope for variety in such areas as the equipment to be carried. This latitude is particularly important in the case of emergency tenders because local conditions and risks vary considerably and make different demands of vehicles and equipment. Emergency tenders therefore probably vary more than other appliances for which JCDD specifications have been prepared.

### b. Types of emergency tender

(1) Type A

JCDD 9 specifies that this emergency tender should be designed for both firefighting and special service work in large cities and ports. Examples of incidents at which it might be deployed are

(i)     large urban fires and other difficult or special fires requiring the use of breathing apparatus, special equipment, or enhanced illumination

(ii)    major electrical fires, involving for example, power stations or transformers;

(iii)   ship fires;

(iv)    lift, road, railway and machinery accidents for which special equipment not locally available is required;

(v)     major leakages or spills of toxic or otherwise dangerous substances.

General features of the Type A emergency tender are described below: those who wish for more detailed information on its design, performance, etc. should consult the specification.

The tender should have an enclosed body, with driving compartment, central or electrical compartment and rear compartment. It should be able to carry eight personnel: the driver and officer in the front and six crew in the rear. Access to the rear compartment should be via a single or double door at the rear, fitted with a dropdown window and a means of holding the door open. The compartment should be large enough for the crew to don b.a. en route and should have adequate natural lighting and ample ventilation. It should be fitted with a bench for servicing b.a. sets, plus internally-lit lockers, shelves, brackets and fittings for carrying the sets and all the other non-electrical equipment of the tender. Electrical equipment should be carried in the central compartment which is occupied principally by a built-in electric generator. This provides current for emergency lighting and portable electric tools. It should have an output of at least five, and preferably, eight kilowatts at 110V., a.c. This limits the maximum voltage to earth to 55V. The generator must be driven by the road engine, preferably by a direct drive from the power take-off. The electrical control panel with its associated protective gear should be on the near side of the central compartment (see Plate 31). Its equipment should all be splash-proof. Other electrical equipment should be in accordance with British Standard Code of Practice 1017 and provide maximum safety. The appliance must have certain accessories, such as a search-light, a spot light, and an inspection lamp; other accessories must be supplied at the purchaser's request. The appliance must be able to satisfy various acceptance tests, covering its road performance, its stability, and the operation of its generator and electrical equipment.

(2) Type B

Like the Type A, the Type B emergency tender is intended for both firefighting and special services, but in smaller towns and ports. It is smaller than the Type A and designed to carry six men only. It is also less comprehensively equipped and in particular it does not incorporate a built-in generator. This is the main distinction between the two types. Lacking the generator, the Type B does not need a central compartment of the Type A sort. In most other ways however it closely resembles the Type A and has to pass the same acceptance tests, excluding those for generator and electrical equipment.

(3) Other types

As noted above, the specifications do allow considerable variation in details of design, equipment etc. Further more as e.ts must depend to such an extent on the local circumstances, Brigades may have requirements for them which are not covered by the specifications.

They may also, for expediency, combine e.ts with other special appliances. There are no specifications for such combinations but the appliances will share some of the features outlined above with some of the features of appliances described elsewhere in this Chapter.

## c. Equipment

The type of equipment carried by an e.t. varies from one to another. Purchasers are required, when issuing invitations to tender, to supply a comprehensive list, with quantities, of the equipment the appliance will have to carry and to state which items firms making a tender should include in their bids. This section refers to some of the more important items of equipment usually carried.

### (1) Lighting equipment

An emergency tender regularly carries various items of lighting equipment (some are required by the specifications). The principal types are described in the *Manual,* Book 2, Chapter 16.

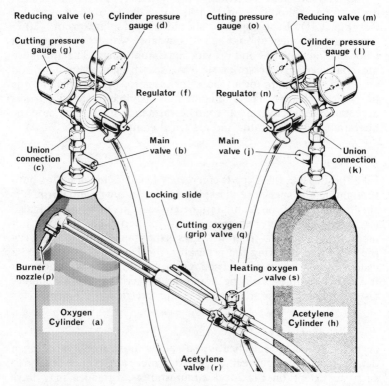

Fig. 14.1 A typical oxy-acetylene cutting plant.

(2) Oxy-acetylene and oxy-propane cutting equipment

Whilst oxy-acetylene equipment is less common now than formerly, it is still in use by a number of Brigades. Fig 14.1 shows a typical example which is capable of cutting through metal up to 150 mm thick. This sort of equipment has considerable useful cutting ability, but is cumbersome and awkward in use. Brigades have therefore been making increasing use of oxy-propane cutting equipment. This consists of a self-contained, portable, lightweight set carried in a harness on the wearer's back. It can be used in any position and has pre-set oxygen and fuel gas regulators. Figs 14.2 and 14.3 show two types of oxy-propane equipment. The first can cut material up to 25 mm thick for ten minutes continuously and the second can cut mild steel up to 50 mm thick and be operated for 15 minutes continuously. Larger capacity oxy-propane sets are also in use: they

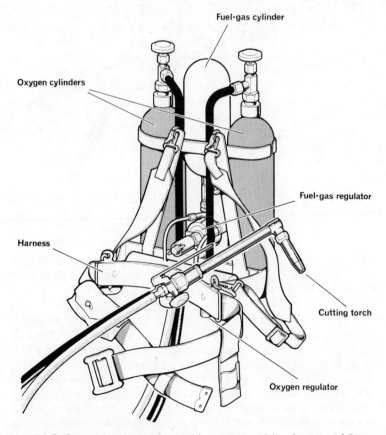

Fig. 14.2 Oxy-propane cutting equipment capable of cutting 25 mm thick material.

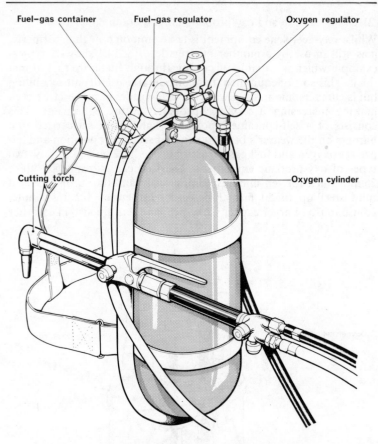

Fuel-gas container      Fuel-gas regulator      Oxygen regulator

Cutting torch      Oxygen cylinder

Fig. 14.3 Oxy-propane cutting equipment capable of cutting material 50mm thick.

are lighter than oxy-acetylene sets because of the comparatively thin walls of the propane cylinder, and they may be considered safer because of the narrower explosive limits of propane.

(3) Hydraulic jacks, air lifting bags and miscellaneous lifting gear

E.ts carry various types of hydraulic jack. All work to the same principle (see Fig. 14.4): the operating lever activates a pump which forces hydraulic oil from a reservoir in the base into the cylinder, thus raising the ram and head. The opening of a relief valve lowers the jack, its speed of descent being controlled by the degree of opening.

Under some circumstances, e.g. where the ground is soft or where awkwardly shaped objects have to be lifted, air bags have an advantage. These are made in different sizes and are usually of reinforced neoprene. They lie very flat when deflated and are

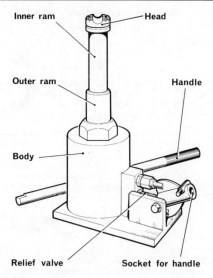

Fig. 14.4 Diagram showing a 5 tonne hydraulic jack.

therefore handy for inserting into small gaps. They lift the object as they are filled with air. Methods of inflation vary. One type requires low pressure only to inflate to over a metre in height and can use an ordinary b.a. cylinder. Other types are inflated to high pressure by a compressor but only expand through about 200–250 mm. The weight the bags can lift varies from three tonnes to over 60 tonnes (see Plate 32).

E.ts also carry chain pulley blocks, lifting chains, double leg chain slings and other miscellaneous gear, including sheer legs.

(4) Hydraulic rescue set and other powered tools

A hydraulic rescue set consists usually of a hand-operated hydraulic pump connected by high-pressure tubing to a ram. It works on the same principle as the hydraulic jack. There are various adaptors and attachments for the ram which enable it to push, pull, open, cut and widen a gap as necessary. Fig 14.5 shows an example of such a set and the *Manual,* Book 12, Chapter 3 illustrates the use of this sort of equipment at road traffic accidents.

Electric tools, such as drills, saws and shears, which are powered by the generator, are also part of an e.t.'s equipment, as are pneumatic tools such as chisels, saws and shears. These can be operated from standard b.a. compressed air cylinders, are self-contained and reasonably portable.

(5) B.a.

Emergency tenders carry a number of b.a. sets and associated equipment (see the *Manual,* Book 6).

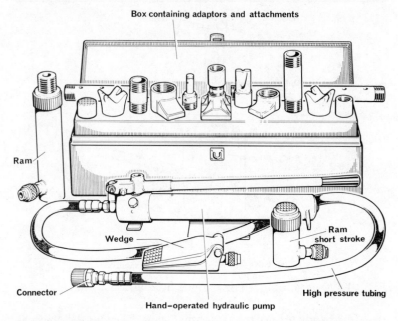

Box containing adaptors and attachments

Ram

Wedge

Connector

Hand—operated hydraulic pump

Ram short stroke

High pressure tubing

Fig. 14.5 Examples of items contained in a hydraulic rescue equipment set.

### (6) Protective clothing

This consists of suits designed to be worn with b.a. and to give additional protection when dangerous chemicals or other such substances are involved in an incident. Other items of protective apparel, such as rubber gloves and thigh boots, are also carried. Owing to the increase in the number of incidents involving dangerous substances many Brigades are now fitting special chemical incident units with this equipment.

### (7) Hand tools

Hand tools carried include much of the small gear described in the *Manual,* Book 2, Chapter 15 — bending bars, rail spreaders, bolt cutters, etc. — together with such items as spades, shovel and large axes. Fig. 14.6 shows a typical set of some of the smaller items.

All equipment should be maintained in accordance with the manufacturer's instructions and, where appropriate, subjected to the standard tests laid down in the *Fire Service Drill Book.*

### d. Examples

#### (1) Type A

Plate 33 illustrates a typical metropolitan Type A e.t. It is fitted with a generator giving 7.5 kilowatt at 110V, d.c. and a compressor

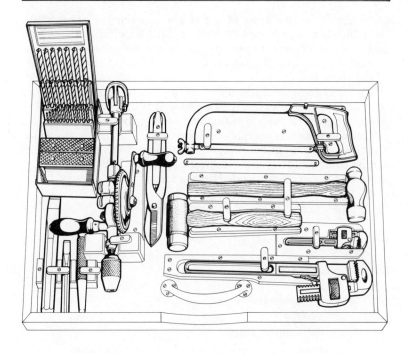

Fig. 14.6 Typical set of small hand tools.

providing up to 0.6 m³/s of air at about seven bar. These have independent hydraulic motors which are fed by a pump driven by a power take-off from the road engine. The control panel and take-off points for both electrical and air supplies are located on the near-side of the appliance.

The air system can supply up to four air tools through 120 m of air line and four 15.2 m air line reels are stowed by the control panel. The generator feeds through an automatic voltage control, meters and a master circuit breaker, supplying four sockets on the control panel and five at the rear of the appliance. There are five cable reels which can be combined to give a total run of 275 m.

Among the items the tender carries are:

(i)      five c.a. b.a. sets with spare cylinders;

(ii)     air lines;

(iii)    control equipment;

(iv)    chemical protection suits;

(v)     decontamination gear;

(vi)    jacks;

(vii)   lifting tackle;

(viii)  Tirfor;

(ix)    cutting gear;

(x)     a wide variety of blocks and tools;

(xi)    special lifting gear supplied by the London Transport Executive.

Fully laden the vehicle weighs 9.4 t. It is moved by a Perkins V8–540 direct injection diesel engine developing 134 kW.

(2) Type B

Plate 34 shows a Type B emergency tender which also functions as a b.a. tender and control unit. It is a good example of a special adaptation for a particular area and risk. The vehicle is a Bedford CFE with a 3.8 m wheelbase, 8.2 litre aspirated diesel road engine, automatic transmission and power steering.

Because of its combined function the vehicle carries additional special equipment such as six c.a. b.a. positive pressure sets with extra cylinders and other equipment necessary to give prolonged support at a b.a. incident, together with an Automan resuscitation set. It also carries additional radio equipment plus a number of personal radios, maps and three 100W halogen flood lamps powered by a 110V portable generator. It has a comprehensive range of usual e.t. equipment: air lifting bags, hydraulic rescue equipment, air-cutting tools supplied by a twin c.a. cylinder portable pack, oxy-propane cutting equipment, Tirfor winch, porto-power jacks, cutting jaws, a decontamination pack, radiation detection equipment, protective gear, etc. Additionally, it carries special items to meet the particular needs of the area it covers. There are lifejackets in view of the long coast-line. The area is littered with old, unfenced mineshafts and the tender is therefore specially fitted to carry gear for rescuing people and animals from such places. It carries on its roof 6.3 m long sheer-legs and in lockers other necessary equipment such as block and tackle, chain tackle, extra long lines (one 230 m x 24 mm), tripod lifting apparatus, pig nets and cattle harness. There is also an adapted helicopter harness and crash helmets to protect firemen being lowered.

A similar combined emergency tender and control unit covers an extensive dock area in a large metropolitan Brigade. This likewise carries specialist equipment peculiar to the risk covered such as a ship's inclinometer, ship stability board, ship thermometer, line portable marine radio, leaky feeder radio equipment and lifejackets.

## 2   Water carriers

In some areas water supplies are short or difficult of access. There are various ways of overcoming this difficulty, such as using a number of water tenders, but many Brigades have experienced problems with setting up and maintaining relays. The water carrier developed to meet this situation. It has several times the capacity of a water tender and can cross rough terrain. Plate 35 shows a typical example, mounted on a Leyland Boxer chassis. It has a tank which can hold 4,500 litres of water and it carries four lengths of 25 m × 45 mm delivery hose, two lengths of 3 m × 100 mm suction hose, normal hydrant gear and a basket strainer. There is also a 1125 l/min light portable pump between the tank and cab. This can:

(i)     pump straight from the tank;

(ii)    lift water and fill the tank via a goose-neck;

(iii)   supply a hose reel.

The water in the tank can also be discharged quickly into a temporary dam.

# 3    Breakdown lorries

A few Brigades still maintain breakdown lorries. They use these not only for the recovery of their own vehicles but also at special services where lifting or towing is required. The lorries usually also carry some of the equipment found on an e.t. and may have some special fitting designed to cope with specific local conditions, such as a snow-plough. The power take-off they use to drive a crane can usually also function as a winch either over the crane boom or as a straight pull.

Plate 36 shows one such appliance, built on a Dennis Maxim chassis and fitted with Holmes 750 twin booms and the slings, sheaves, tackle and other ancillary equipment for use with them. The booms, which are driven by a power take-off or can be operated manually, can lift 7.5 tonne. Hydraulic jacks can be used for additional stability. The appliance carries an inbuilt compressor to supply rescue equipment, jacks, air-bags, blocks, cutting equipment, warning signs and additional portable lighting.

# 4    Decontamination/Chemical incident unit

The Introduction to this Part of Book 5 mentioned that the increasing use and transportation of dangerous goods are one particular source of problems for the Fire Service. When incidents occur there is a need for special procedures, particularly to cope with the decontamination of those who may have been in contact with the dangerous substance. The *Manual,* Book 12, Part 2 discusses this whole subject.

A number of Brigades have now equipped themselves with special decontamination/chemical incident units to attend such incidents as necessary. Plates 37 and 38 show a particular example mounted on a Bedford TK 860KD chassis with a Bedford 330 six cylinder diesel engine of 73 kW. The whole unit comprises two sections plus a collapsible canopy which can be put up at the rear. Those requiring decontamination are first cleansed and stripped of protective clothing in a decontamination zone outside the unit then enter under the canopy and undress. They shower in the rear compartment, shower in the front compartment, dry themselves, then dress in clean gear and exit from the front.

The vehicle carries two 20 kg l.p.g. cylinders which supply two flameless wall heaters and one water heater. To illuminate the area it also has an extendable floodlight mounted on its roof. This is supplied by an integral 2.5 kW Powermite alternator from the road engine. Four 200 bar air cylinders feed the decontamination team's air-line equipment, which consists of 30 m of 7 mm tubing on a drum. The remaining general equipment of the unit is similar to that listed in the *Manual,* Book 12, Chapter 8, section 3d, plus several types of instrument for monitoring radiation.

## 5    Foam tenders

Foam tenders vary considerably in size, capacity and equipment but all have the same main purpose of delivering foam concentrate to the fireground in bulk. The simplest type merely backs up other appliances by carrying concentrate and equipment to an incident for them to use. Other foam tenders take the form of tankers with a minimum pumping capacity. Some sophisticated appliances have fixed foam monitors and equipment and capacities akin to those of airport crash tenders. The type a Brigade might select will take into account the risks of the particular area.

Plate 39 shows a foam tender used in an area where there is a limited risk of incidents where foam in bulk might be needed. The vehicle, which is powered by a 185 kW engine on a Dennis chassis, carries 910 1 of standard protein foam concentrate and 1800 1 of water. It has two pumps driven by power take-offs from the road engine: a 200 1/min. Albion positive pressure pump for the foam concentrate and a 4,500 1/min. Dennis No 3 water pump. The tender's capabilities are:

(i)     to transfer foam concentrate from its tank to another vehicle or dam;

(ii)    to meter foam concentrate from an external source to its own tank, another vehicle or dam;

(iii)   to meter foam concentrate into the water supply for branches or monitors working from itself;

(iv)    to produce foam from its two No 10 generators;

(v)     to act as an ordinary high capacity pump.

The tender also carries 225 1 of high expansion foam concentrate to supply an h.e.f. unit and has facilities for carrying a Jetmaster monitor and foam branches of various sizes. Two dams, of 225 1 and 450 1 capacity respectively, can hold the necessary open volume of foam concentrate for the Jetmaster monitor and other appliances.

## 6    Rolonoff units, pods, caravans, etc.

As mentioned in the Introduction, a number of Brigades, to economise on running costs, are now experimenting with means of getting special equipment etc. to an incident without devoting a vehicle wholly, exclusively and permanently to that role. Their means of doing this is to have a single 'prime mover' which can carry or tow any of a selection of special units to where it is required, leave it there, and return to base for another call. Such a prime mover can also transport non-operational units such as a b.a. smoke training chamber. The units themselves, whatever their purpose, can take various forms. This Section describes three examples of what appears to be an increasingly common method of improving Fire Service mobility.

## a. Types

(1) Rolonoff unit

Plate 41 shows a Dodge G13 chassis fitted with a fully hydraulic demountable body system which operates three hydraulic rams and a central lifting arm, or goose-neck, which terminates in a hook. The system can lift 40 t. The driver steers his unladen vehicle under the body runners and draws the goose-neck forward, pulling the unit onto the frame where automatic locks secure it.

(2) Pod

The prime mover shown in Plate 42 is on a Bedford TK 12/60 chassis. It carries a demountable pod with a maximum weight capacity of four tonnes. The pod has four retractable legs operated electro-hydraulically from an outside electrical supply or by batteries; on other types of pod they can be hydraulically hand-pumped into position. When the legs are retracted the pod rests with its floor about 150 mm above the ground; when they are extended it is free-standing. To load, the prime mover is backed under the raised pod, which is set correctly into place by guides on its underside and secured to the chassis by twist locks.

(3) Caravan

Plate 43 shows the adaptation of a 6.4 m by 2.2 m, four-wheeled caravan as a specialist unit which can be towed into place. The unit shown is a decontamination unit but the same Brigade also uses this method to provide a control unit and a b.a. tender. Fully loaded the caravan weighs 1.75 t.

## b. Operational considerations

Usually once a unit of this type has been demounted it cannot be moved without its prime mover. Careful consideration must therefore be given to its positioning at an incident, particularly in the case of a decontamination unit. The following points are particularly important.

(i)   obviously, the unit must not block access to and from the incident;

(ii)    in most cases it must be up-wind of the incident; this is particularly important with decontamination and b.a. units. Ensuring this can be a problem on a relatively still day or on a day when the wind is constantly changing direction;

(iii)   care should be taken about the state of the ground on which the unit is standing, if it is off the highway. That which gives support at the start might become flooded by water run-off as the incident progresses. A foam tanker unit, especially one supplying a dam or receiving water, could become bogged down;

(iv)    in the case of control units there are radio 'dead' areas to be avoided. Even a movement of a few metres can bring about substantial improvements in reception and sending of messages;

(v)     access for the prime mover to come and remove the unit must be kept open.


## 7   Control units

Any large incident requires a central point from which the officer in charge can co-ordinate operations. This usually consists of an appliance fitted out as a mobile administrative and communications centre and called a control unit (see Plate 45). Its main purpose is to give the officer in charge a centre where he can send and receive information, keep in touch with his Brigade control, plan his tactics and muster and deploy his resources. For ease of identification the unit is marked with red and white checks round its side; it usually also has an internally lit translucent dome, similarly chequered, or a flashing beacon on a telescopic mast on its roof.

The unit must be comprehensively equipped. Details will vary according to individual Brigade requirements but the following items are usually present:

(i)     a V.H.F. radio transmitter/receiver for communication with Brigade control;

(ii)    a fireground radio system consisting of a number of small portable transceivers with a separate control and operator in the unit;

(iii)   large-scale maps of the area with details of water supplies, high risks, access, etc.;

(iv)    a means of displaying representationally the incident area and position of crews, appliances, b.a. controls, equipment dumps, etc.;

(v)    a tape-recorder for messages in and out;

(vi)    a small library containing information on orders and procedures, essential non-Brigade resources, chemicals, high and low tides, etc.;

(vii)    emergency lighting, accurate timepieces, field telephones, portable generator;

(viii)    a rack for nominal roll boards and check list of appliances and officers in attendance.

As well as their main unit some Brigades also have smaller appliances to use as controls at smaller incidents. These are sometimes called divisional control units. They carry the same equipment to a lesser degree (see Plate 44).

## b. Operational considerations

The use of a control unit at an incident, its siting, etc. are discussed in the *Manual,* Books 11 and 12. Book 10 discusses communications.

## c. Examples

As already explained, many special appliances serve two or even more purposes and control units can be based on vehicles having some other function too; a combined control unit/emergency tender/breathing apparatus tender has been described in section 1d above. The two examples that follow are however used solely as control units.

### (1) Control unit (metropolitan Brigade)

Plate 45 shows an appliance based on a Ford R1014 bus chassis. It has three sections: a communications area, a working area and a conference area. It is equipped with air operated radio masts, on one of which are mounted twin 500 W floodlights, and an Orion 6 kW generator. The radio system is independent and battery powered but the generator can take over the supply; it can also keep the battery topped up. For prolonged incidents the unit also carries a 30 m, 28 amp mains extension cable which the local Electricity Board can connect into a domestic supply. To help prevent accidents a light on the driver's panel shows when the masts are raised and an alarm sounds if the handbrake is released.

The unit has the following communication facilities (see Plate 46):

(i)     a main radio which can be tuned to any of the five Brigade channels and which has two generator positions;

(ii)    a portable Westminster set which can be used at a forward control point and which provides instant communication with the control unit and any of the Brigade controls;

(iii)   13 three channel portable EVAC personal radios (channel A: appliance fireground communications; channel B: messages to the control unit; channel C: communication between senior officers);

(iv)    an EVAC channel C simplex transmitter/receiver with a switching arrangement to use leaky, or radiating, feeder for underground communications. The feeder cable socket is on the near-side of the vehicle;

(v)     an EVAC channel B master set transmitter/receiver;

(vi)    an EVAC channel A/C transmitter/receiver;

(vii)   a 55 channel radiophone fitted with one telephone for the operator and one for senior officers;

(viii)  Figaro equipment, consisting of a portable base transmitter/receiver and a surcoat with transmitter/receiver and combined headset/throat microphone;

(ix)    field telephones of two types: one consisting of ex-military sets with cable reel, the other of a small switchboard in the unit two cable drums marked 'Police' and 'Ambulance' respectively;

(x)     two sets of Diktron b.a. communications equipment.

The working area has flat working surfaces, map drawers, information files and a board giving at-a-glance information on appliances attending, crew position, fireground, etc. The conference area consists of a small room at the rear where senior officers of the brigade and other emergency services can gather to discuss operations.


## (2) Control unit (shire county Brigade)

Plate 47 shows a unit based on a Dodge 50 chassis with a Perkins 4 litre diesel road engine. It has two parts. The radio compartment contains the following facilities:

(i)     a main Brigade radio scheme set;

(ii)    a controlling position for a portable network of eight Burndept personal radios linked via a W150 base station, entirely independent of the main scheme;

(iii)    provision for a British Telecom line to be connected via an external jack point.

In the other part of the unit are mounted a large county map and perspex-covered mobilising board (Plate 48). There are also perspex-covered working surfaces which are hinged to allow the insertion of 1: 10,000 sectional maps or the use of a white underlay for rough sketches of the incident area etc. Under the working surfaces are map drawers and files of other information. The compact arrangement allows permanent seating for three and four other chairs in the 19 m³ space.

## 8    Hose-laying lorries

There are many areas, particularly in rural districts, where water supplies are limited or, sometimes, non-existent. Even when the supply is sufficient for normal use a large fire may make extra demands. It would be very time-consuming in such circumstances to lay out by hand a double line of hose from the nearest good supply, perhaps a kilometre away, to the fireground. To overcome this problem some Brigades employ a hose-laying lorry. This is designed to carry 90 mm hose, flaked and coupled in continuous lines along its length, and to lay it out in single and double lines at speeds up to 50 k.p.h. Plate 49 shows a typical example, built on a Ford 9.5 tonne D series chassis powered by a 4.2 litre diesel engine. This carries 80 lengths of coupled 90 mm hose: enough for 1,850 m of single line or twin 925 m lines simultaneously laid.

# Part 6
# Pumping appliances

## Introduction

This last Part of Book 5 looks at perhaps the most basic of all Fire Service appliances, the pumping appliances. 'Pumping appliance' is a generic term applied to a variety of the Brigade vehicles which form part of the first attendance at all incidents. They may carry different sorts of sophisticated equipment for dealing with special services as well as with fires but the prime function of all is to carry personnel and their basic equipment to an incident and to pump water at the scene. All are usually equipped with a centrifugal water pump, hose, ladders, and small gear.

Pumping appliances have developed considerably over the last 20 years. As already noted, they have become increasingly likely to carry a 13.5 m extension ladder rather than an escape; other examples of change are an increase in the amount of water carried and the up-grading of pumps to a multi-purpose role. As the number of types of appliance has increased so have the names given them: pump escape, pump, dual purpose appliance, water tender escape, water tender ladder, pump ladder, L4 pump, are just some examples.

This part of Book 5 looks at the general features of pumping appliances, gives a few particular examples and discusses the principles of working with them. Pumps themselves and pump operation are described in the *Manual,* Part 2.

# Chapter 16
# Pumping appliances

## 1   JCDD specifications

Three specifications apply to pumping appliances:

(i)     JCDD 3/1    water tender, Type B (JCDD 3 for a water tender, Type A, i.e. one not incorporating a built-in pump, has been withdrawn);

(ii)    JCDD 4      dual purpose appliance/i.e. an appliance carrying either a pump and an extension ladder, known as a pump, or a pump and escape, known as a pump escape);

(iii)   JCDD 18     2,270 1/min pump appliance.

The specifications were all promulgated in 1975 and amended in 1980. They are very comprehensive and should be consulted by those who want more precise details about the appliances. Among the requirements they lay down are:

(i)     an engine capable of developing at least 12.1 kW per tonne of the gross vehicle weight;

(ii)    an engine cooling system permitting continuous stationery running without overheating when the power take off only is engaged;

(iii)    a centrifugal pump, preferably single stage only, having set performance ratings including the ability to operate at both high and low pressure;

(iv)    accommodation for at least six personnel;

(v)     the ability to carry safely various sizes of ladder;

(vi)    even weight distribution;

(vii)   acceleration, top speed, braking and hill-start capabilities to set standards;

(viii)  compliance with current *Motor Vehicles (Construction and Use)* and *Road Vehicles (Lighting) Regulations.*

The specifications also lay down tank capacities. Thus, a water tender should be capable of carrying between 1,8201 and 2,2701 of water, a dual purpose appliance should have a minimum capacity of 4551 and a 2,2701/min pumping appliance a minimum capacity of 3651. In fact, however, Brigades are currently tending to favour 1,3551 tanks; usually only the smallest appliance, based on Land Rovers or adopted from 1½–3 tonne vehicles, carry the minimum quantities.

## 2    Pumps

### a. Types

Obviously Brigades will make different demands on their pumping appliances. There are therefore various sizes and types of pump fitted. Those commonly found have capacities of 1,3551/min, 2,2701/min, 3,5001/min and 4,5001/min (all at about seven bar pressure); there is additionally the multi-pressure pump which can supply 2251/min at between 30 and 40 bar.

### b. Pump mountings

The position of a pump on an appliance depends on a number of factors such as whether the appliance carries an escape, its chassis size, the equipment it carries and the locker facilities required. The three mountings possible are described below.

#### (1) Rear mounting

This is really only suitable for an appliance not carrying an escape, because of the obvious drawback of having to slip the escape before being able to operate the pump. The arrangement gives maximum locker space and is particularly suited to chassis with a short wheel base. It does however require lengthy p.t.o. shafts which reduce efficiency and may cause maintenance problems. Mounting the pump at the rear but having the controls at the side overcomes some of these problems but long p.t.o shafts remain and either the appliance has to be longer or the amount of locker space has to be less. Access to equipment on top of the appliance may also be obstructed (see Plates 50 and 57).

#### (2) Midships mounting

This has the advantages that an escape need not be slipped before pump operation, that controls can be duplicated to allow operation from either side, and that p.t.o shafts are shortened, giving increased efficiency. The serious disadvantage of this mounting is the pump's lack of accessibility for maintenance, etc.

#### (3) Front mounting

This arrangement is common in the USA but has never been popular in the UK. It is found occasionally on Land Rover adaptations. The drive is direct from the crankshaft and usually requires some sort of gearing to give adequate pump revolutions without running the engine at excessive speed (see Plate 51).

# 3   Pump power units and drives

Pumps have to be driven by an external power unit, or engine, and their design must relate to the power it develops. Manufacturers' specifications will quote this as the engine's brake horse-power, measured in b.h.p. or, under metrication, simply as its brake power, measured in kilowatts (eg 225 b.h.p/190kW). On Fire Brigade appliances pumps may be either 'built-in pumps', driven by the vehicle's road engine or portable pumps, coupled to an independent engine to make a complete portable unit.

### a.  Power take-offs (P.t.os)

For a vehicle's road engine to drive a built-in pump its power must be diverted from driving the road wheels. This is effected by a power take-off. (Similar systems can be used to drive the hydraulic pumps which power turntable ladders and hydraulic platforms: see Chapters 4 and 11). There are several types of p.t.o for driving fire pumps:

(i)     from the gear-box;

(ii)    from the drive shaft to the gear-box;

(iii)   from a transfer box

Fig. 16.1 shows these diagrammatically.

(1) Power take-off from the gear-box

This is usually taken from the main shaft in the gear-box (Fig. 15.1 (1)) and because of its position is known as a 'side' or 'window' p.t.o. There will normally be a step-up gear between the main shaft and the p.t.o. This type of p.t.o has a limit on power output.

(2) Power take-off from the drive shaft to the gear-box

This type (Fig 15.1 (2)) is geared to the drive shaft between the engine and the gear-box, being known therefore as a 'sandwich' p.t.o. (see Plate 52). It likewise incorporates a step-up gear between the drive shaft and the p.t.o.

(3) Power take-off from a transfer box

In this method (Fig 15.1 (3)) the output shaft from the gear-box drives a transfer box. This in turn drives the rear (and often the front) axle or the fire pump as required, the latter sometimes through a side p.t.o. A single lever usually engages window and sandwich p.t.os but with a transfer box it is necessary first to engage the pump drive in the transfer box, then the appropriate gear in the vehicle gear-box. In all cases manufacturers' instructions must be followed.

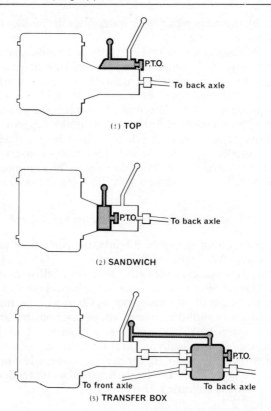

Fig. 15.1 Diagram showing types of power take-off (1) from the gear box (2) from the drive shaft to the gearbox (c) from a transfer box.

(4) Automatic transmission

Both window and sandwich p.t.os may be used with automatic transmission but the former has limits on power output. With automatic transmission it may be necessary to select an appropriate drive range before engaging the p.t.o. This will again depend on manufacturers' instructions.

## 4 Hose reels

All normal pumping appliances carry one hose-reel and the majority two, usually one either side of the appliance. Each reel should carry at least 55 m of 19 mm internal bore hose, its construction depending on whether it is for high or low pressure use. It should have a potential output of at least 2251/min at 24 bar. JCDD 25 specifies standards that the hose reel branch should attain in such areas as output, leakage and adjustability (to give a jet, spray or fog). The hose reels are supplied by the main pump with water from the tank or from an external source (see Plate 54).

The main function of a hose reel is to provide an immediate supply of water for extinguishing small fires and for making an initial attack to hold larger fires in check whilst main jets are got to work. Hose reels are light, manoeuvrable and easily controllable for damping down and turning over. An additional advantage is the possibility of detaching the hose from the reel and, with a suitable adaptor, using it to extend a line of hose or, with a dividing breeching and adaptors, splitting it to supply two jets. The use of hose reels conserves water and applied intelligently, cuts water damage to the minimum, since all hose-reel nozzles can produce a spray or fog and can be shut off.

## 5  Standard equipment and storage

Obviously, even within the same Brigade, the equipment pumping appliances carry will vary according to such factors as local requirements, preferences, experience and availability. Books 2, 3 and 6 of the *Manual* described the items all appliances normally carry, principally: c.a.b.a. and the necessary auxiliary equipment for its efficient use; delivery and suction hose, couplings, branches and other hose fittings; ropes and lines; small gear. Chapter 1 of Part 6b mentions gear used specifically in rural firefighting.

However, growing demands on the Service have made it necessary to introduce extra items. Accordingly a typical water tender ladder might now carry the following also:

(i)     chemical protection suits;

(ii)    a chemical incident board;

(iii)   information on hazardous chemicals;

(iv)    fluorescent surcoats;

(v)     goggles, masks and ear muffs;

(vi)    a 'Police Accident' sign;

(vii)   radiation monitoring equipment;

(viii)  a Tirfor winch;

(ix)    compressed air outlets and air-lines;

(x)     airbags;

(xi)    a generator;

(xii)   portable radios.

The *Manual,* Book 12, mentions in addition the various hydraulic/pneumatic cutting and spreading tools which it is recommended at least some appliances should carry. More powerful versions of these are now common.

Book 12 also reflects the greater emphasis being given to handling casualties and to decontamination. To help in casualty handling, appliances in general may now carry some sophisticated resuscitation and first aid equipment. For decontamination some Brigades find it desirable for certain appliances to carry a special decontamination pack; this facilitates initial procedures before the arrival, if necessary, of a chemical incident unit (see above).

To accommodate all this equipment without increasing appliance size requires ingenuity and careful planning. Plates 55 and 56 show examples of stowage.

## 6   Examples of appliances

As already noted, there are many different pumping appliances in use by UK Fire Brigades and it is only possible to cover relatively few here. This Section gives however a fairly representative sample.

### a. Dual purpose appliance

One particular type, supplied by Cheshire Fire Engineering is based on a Shelvoke and Drewry Type WX chassis, powered by a Perkins V8 640 engine (see Plate 57) has a rear-mounted pump with side controls, and can carry 1,365 litres of water. It can be fitted with an escape, a 13.5 m extension ladder or a nine metre extension ladder and can carry all the standard gear including other ladders. Its unladen weight is 8.25 t.

#### (1) Engine and transmission

The Perkins V8 640 is a direct injection diesel type engine which develops 162 kW at 2,600 r.p.m. with a maximum torque of 655 Nm at 1,650 r.p.m. A gauge in the driver's cab keeps a check on engine running hours. Transmission is by Allison MT 643 fully automatic gearbox with four forward and one reverse gear, torque convertor and automatic lock up. This should give a maximum speed of about 90 K.p.h.

#### (2) Fire pump

The appliance carries a two-stage multi-pressure Godiva pump with a maximum output of 4,5001 at 5.5 bar or 2051 at 54 bar. It has a manual primer of the water ring type and a separate priming tank to fill it. The power take-off is engaged hydraulically and there are safety interlocks to prevent inadvertent operation whch could cause damage. A control on the side panel changes the water delivery from low to high pressure and there are separate gauges for the different pressures in addition to the usual compound gauge and tachometer (see Plate 50).

## (3) Hose reel

Two hose reels each carrying 55 m of high pressure tubing are fitted in lockers midships. The hose reel control valves are fitted in the feed supply pipes and are operated from the near-side pump control panel. A direct reading gauge adjacent to the control panel indicates the contents of the tank. The hose reels can be operated at either low or high pressure, either separately or both at the same time.

## (4) Compressed air facility

The braking system provides for three separate air systems compressed, by an engine driven compressor, to nine bar. It is possible to couple up two 15 m air-lines to connections at a panel on the near-side and use them simultaneously to supply r.t.a equipment etc. The engine has to be kept running at about 1,500 r.p.m. and, if the pump is not in use, the gear selector must be in neutral.

## b. Water tender ladders

JCDD 3/1 lays down a detailed specification for a water tender with built-in pump (JCDD 3 for a water tender towing a trailer pump has been withdrawn). The appliance must be designed for both urban and rural use and to carry an extension ladder or escape, a light self-contained portable fire pump conforming to JCDD 30 and two hose reels. Its water tank should hold 1,825–2,2751 and the minimum output of the pump should be 2,2701/min at seven bar. The appliance can be two or four-wheel drive and should accommodate a crew of six.

## (1) Conventional

Whilst this type of appliance generally follows the JCDD specification referred to above there is great variation from Brigade to Brigade to reflect particular local needs.

Plate 58 shows a typical example from the Dennis R133 series. It carries a rear- mounted pump with rear controls, an 1,8001 water tank and 13.5 m ladder. Amongst its other features are the following:

| | | |
|---|---|---|
| (i) | engine and transmission | the engine is a V8–640 diesel, developing 162 kW at 2,800 r.p.m with transmission by an Allison MT643 four-speed gearbox; the cooling system is pressurised, thermostatically controlled, and closed circuit. There is a sandwich-type p.t.o.; |
| (ii) | fire pump | this is a two-stage, multi-pressure, UMP pump with a maximum output of 4,5001/min at 5.5 bar. The primer is a water ring type fitted with a header tank. There is also, on this vehicle, a round-the-pump variable inductor with a manual control |

|          |           | setting incorporated into the pump system; |
|----------|-----------|--------------------------------------------|
| (iii)    | hose reels | two hose reels, each with three 20 m lengths of high pressure 19 mm hose, are carried on either side behind the rear road wheels; |
| (iv)     | equipment  | besides the 13.5 m ladder the appliance has roof gantries to carry a three section 4.5 m ladder and a roof ladder. Also on the roof are two cable reels each with 90 m of cable to supply two searchlights. In racks at the rear of the appliance are six 2.5 m lengths of 140 mm suction hose. These are used with the UMP pump and by means of a 100 mm to 140 mm adaptor can also serve the light portable pump carried in the nearside front locker. Compressed air breathing apparatus, chemical protection suits, air operated and hydraulic r.t.a equipment, Tirfor, foam concentrate, an FB5X branch, and the normal complement of hose and small gear, complete the equipment. |

The appliance usually has a crew of five.

## (2) Alternative conventional

Plate 59 shows another example of a conventional water tender ladder. This is a design by HCB-Angus on a 3.8 m wheelbase also with a rear-mounted pump an 1,800 l water tank and a 13.5 m ladder. Amongst its other features are the following:

|        |                        |                                                                                 |
|--------|------------------------|---------------------------------------------------------------------------------|
| (i)    | engine and transmission | the engine is a Bedford 500 diesel rated at 120 kW and the transmission is by an Allison four-speed manual gear box with a sandwich p.t.o. This drives a Godiva UMP 50A multi-pressure pump with three deliveries. The pump can produce either 2250 l/min at seven bar or 200 l/min at 25 bar. An automatic water-ring primer with a separate header tank is fitted; |
| (ii)   | hose reels             | there are hose-reels on either side of the back of the appliance, behind the rear wheels; |
| (iii)  | equipment              | in addition to the 13.5 m ladder, there are on the roof, a three-piece short extension |

ladder and a roof ladder. There are also a searchlight with a 280 m cable reel and a Stemlite telescopic lighting mast which can extend to 2.5 m above vehicle height. An Angus 1,200 1.w.p giving 1,4501/min at seven bar is stowed in the nearside front locker.

Apart from equipment similar to that mentioned in (1) above, the appliance also carries such items as an aircraft axe, motorway maps and hay forks. These reflect the slightly different types of risk the appliance covers.

## (3) Compact

Many Brigades have experimented with what are generally known as 'compact' appliances. There are usually smaller and lighter, and in some cases, faster than the conventional water tender/water tender ladder. They have been developed in response to various local conditions, e.g. narrow lanes, rough, hilly terrain, inner city and old areas with congested streets, podium and pedestrianised areas. They need careful designing to keep their weight down and yet carry sufficient equipment and personnel to be efficient.

Plate 60 shows an example based on a Ford 'A' series chassis with a 3.3 m wheelbase and a 3.5 litre six-cylinder diesel engine. The appliance carries a rear mounted Godiva 2,2701/min UMP multi-pressure pump and a water tank holding 675 litres. It has hose-reels behind the rear wheels on each side, each reel carrying 55 m of high pressure tubing and a hand-controlled nozzle. On the roof are a 10.5 m ladder, a three-piece short extension ladder and a roof ladder. The remaining equipment consists of 16 lengths of 70 mm hose, three c.a.b.a. sets and the usual small gear. The appliance has a crew of five and its fully laden weight is 5.8 tonne.

## c. Simonitors

These appliances are designed to produce and deliver large quantities of foam from a height. They are especially useful at incidents in oil refineries, for delivering foam onto tanks or for cooling purposes. There is no JCDD specification.

Plate 61 shows an example on a Dennis chassis with a 175 kW Rolls Royce petrol engine. The following are its main features:

(i)    a turntable and boom, a water pump and a foam pump. The boom can extend to 12.6 m and operate from − 10° to 90°. The water pump is a Coventry Climax 4,0001/min type with a water-ring primer, four deliveries and a five-way collecting head. The foam pump is a 1821/min rotary gear;

(ii)    a tank containing 1,800l of foam concentrate, an h.e.f. unit and 135l of h.e.f. concentrate;

(iii)   separate power take-offs for the turntable boom and jacks, the water pump and the foam pump;

(iv)    two hydraulically operated outrigger jacks fitted just behind the rear wheels.

For correct operation at least two water tenders are needed to provide the necessary pressurised supply. Water requirements vary according to the use of the boom, e.g. to produce foam, or to give a cooling jet. Other foam or cooling jets can also be supplied.

## 7   Working with pumping appliances

As with all Fire Brigade operations successful working with pumping appliances depends to a large extent on the lessons learnt from practical experience. This Section briefly mentions some of the factors firemen should bear in mind. Factors relating to the pump itself, rather than to the appliance on which it is mounted, are dealt with in the *Manual,* Part 2.

### a.   Before an incident and getting to work

A driver taking over an appliance should follow a set routine. This should invariably include checking the petrol or fuel oil level in the tank, checking the oil level in the sump and seeing that the radiator or head tank is full where appropriate. He should perform similar checks on any portable pumps carried.

Before taking an appliance out of a station the driver should ensure that any immersion heater is unplugged or heating device removed. On arrival at an incident he should choose his site carefully and firmly apply the hand brake. The following are among considerations he should bear in mind:

(i)     when working from a pressure fed supply he should park as close to the fire as possible. This keeps delivery lines short and gives better control and communications;

(ii)    when working from open water he should park as close to the source as possible. This keeps the length of hard suction shorter, cutting down friction loss and the time taken to get to work;

(iii)   he should keep close to the side of the road. Although normal traffic may be stopped, other emergency vehicles may require access;

(iv)    he should ensure the ground is firm enough for the appliance, remembering it may become waterlogged through firefighting operations.

## b.  General considerations in running an appliance; appliance faults

Starting an appliance from cold presents few problems where operation of the choke is automatic. Where operation is manual the driver must return the choke to its normal position as soon as the engine is running smoothly. Failure to do so will lead to sooting up of the sparking plugs on petrol engines and dilution of the lubricating oil. The driver must also avoid frequent stopping and starting before the engine warms up as this leads to excessive wear. He should be familiar with the normal working oil pressure of the appliance and maintain this during operations, topping up the sump if necessary. He should be aware that when there is a sudden or substantial drop in oil pressure he must stop the engine immediately to prevent a complete mechanical breakdown. He should watch the appliance's fuel consumption at all times and inform his officer in charge when he needs more supplies. Drivers should be fully conversant with any peculiarities of their appliance and take them into consideration during operations. Although problems sometimes arise from inexperienced handling, regular servicing and testing should minimise mechanical problems on the fireground.

## c.  Maintenance

Appliances should be serviced and maintained as indicated by the manufacturer and in accordance with Brigade procedure. This should include checking the ladder mountings and fitments and testing pumps and hose reel equipment as laid down in the *Fire Service Drill Book* (see also *Manual,* Part 2). On return from an incident, petrol/fuel oil, lubricating oil and water should be checked. Where it has not been possible at the incident the water tank should be replenished and if other than fresh water has been used, the pump system, tank, hose reels and suction should be thoroughly flushed through with fresh water.

In winter anti-freeze should be added to the cooling water of appliances. If there is any danger of freezing on the way to or from an incident the pump casings and hose reels should be drained. The hose reel tank however should not need special precautions. Immersion heaters and other warming devices should be used on the appliance as necessary.

# Glossary

This glossary consolidates, summarises, and in places expands on information given in the text. Its purpose is four-fold:

(i)     to explain expressions which are possibly unfamiliar;

(ii)    to explain familiar expressions used in an unfamiliar way;

(iii)   to give precise definitions to expressions sometimes used more generally;

(iv)    to spell out abbreviations.

**Air lifting bag**
An inflatable bag, usually of reinforced neoprene, which is used to lift or separate objects as it expands.

**Angle iron**
An L-shaped metal bar, usually of mild steel, used in light construction work.

**Appliance**
Any Fire Service operational vehicle.

**Axle lock**
A device on some turntable ladders to stop the chassis rising on the road springs at one side when the ladder is extended on the other.

**B.a./B.a.t.**
Breathing apparatus/Breathing apparatus tender.

**Ball race**
Ring marking the track in which a ball bearing travels; the ball bearing itself.

**Beaufort Scale**
Numerical scale of wind speeds, with a description of the wind and its effects. Higher numbers indicate increasingly higher speeds (Force 8 indicates a gale).

**B.L.**
Breakdown lorry.

**Bobbin line**
A light, 40 m long line usually made of plaited cotton cord and wound on a leather reel, or bobbin, carried in a pouch on some hook ladder belts and used for hauling up.

**Boom**
One of the moveable, jointed sections of a hydraulic platform, to the topmost of which the cage is fixed.

**Bridge**
To use a ladder or ladders at or near the horizontal in order to cross an open space.

**Burgee**
A small flag.

**C.a.b.a.**
Compressed air breathing apparatus.

**Cage**
The personnel carrying compartment fitted to the topmost boom of a hydraulic platform or turntable ladder.

**Camber**
A slope of the ground from one side of an appliance to the other.

**C.F.B.A.C.**
Central Fire Brigades Advisory Council.

**Chemical incident unit**
An appliance specially equipped for dealing with a chemical incident and carrying out decontamination.

**C.i.u.**
Chemical incident unit.

**Control interlocks**
Devices on a turntable ladder which stop movement of the ladder until the jacks are down and prevent retraction of the jacks until the ladder is fully housed and depressed onto the head-rest.

**Constant speed button**
A button which brings the control levers of a turntable ladder into operation at the correct pressure and stops the movement, bringing levers back to neutral, if released.

**Control unit**
A mobile administrative and communications centre for the officer in charge of an incident.

**C.u.**
Control unit

**Dead man's button/handle/lever**
A device which must be kept constantly engaged for an appliance to be operable.

**Decontamination**
The process of removing dangerous substances from a person's body, clothing or equipment.

**Deflection test**
A test of a ladder's ability to return to its normal position after being stressed by the application of a weight.

**Depress**
To lower the head of a turntable ladder by decreasing the angle of elevation or to reduce the height from the ground of a hydraulic platform's cage.

**D.p.**
Dual purpose appliance.

**Dual purpose appliance**
An appliance designed to carry a fire pump and an escape or extension ladder. It carries a limited amount of water in a tank.

**Elevate**
To raise the head of a turntable ladder by increasing the angle of elevation or to increase the height from the ground of a hydraulic platform's cage.

**Emergency tender**
An appliance specifically designed to carry a wide range of equipment for use in both firefighting and special service work.

**Epicyclic gear**
System of gears in which one or more wheels travels round the outside or inside of another wheel of which the axis is fixed.

**Escape**
A ladder mounted on wheels.

**E.t.**
Emergency tender.

**Extend**
To run out the extending parts of a ladder, increasing its length.

**Extend to lower**
To extend a ladder in order to clear the pawls for lowering.

**Fairlead**
A guide controlling the running of a line.

**Field of operations indicator**
Panel at a t.l.'s controls showing the position of the ladder, limits reached, operations effected, etc.

**Foam**
The expanded mass of bubbles which results from mixing foam concentrate with water and air.

**Foam concentrate**
A manufacturers' product which produces foam when mixed with water and air.

**Foam solution**
Foam concentrate dissolved in water.

**Foam tender**
A special appliance carrying bulk foam concentrate and usually equipped to pump prepared foam onto an incident.

**Footing**
Steadying a ladder by placing one foot on the lowest round or jack beam, grasping the strings, and pressing on the ladder.

**Fo.T.**
Foam tender

**Fulcrum frame**
The structure mounted on the turntable of a t.l. or h.p. whch supports the ladder sections or booms and houses the operating mechanism.

**Gallows**
The headrest for the sections of a t.l. when they are fully depressed onto the appliance.

**Gimbal**
A device to permit a piece of equipment free movement in different directions or to keep it horizontal irrespective of other moves that might affect it.

**Gooseneck**
A U-shaped fitting.

**Gradient**
A slope of the ground from front to back of an appliance.

**Guy line**
A line used to help maintain the stability of a t.l. or to pull on another line being used for lowering so as to keep the person or item being lowered away from obstacles.

**G.v.w.**
Gross vehicle weight.

**Heel**
The foot of a ladder.

**Heel in/out**
To move the foot of a ladder towards/away from a wall.

**H.e.f.**
High expansion foam.

**Height (of an h.p.)**
This distance of the cage bottom from the ground.

**Hermetically**
So as to be air-tight.

**H.L.L.**
Hose laying lorry.

**Hook ladder**
A short ladder with a hook at one end by which it can be suspended for climbing.

**Hook ladder belt**

A belt specially designed for use with a hook ladder. It has a clip for hooking the wearer onto the ladder and a bobbin pouch.

**H.p.**

Hydraulic platform.

**House**

To retract the extensions of a turntable ladder.

**Hydraulic platform**

An appliance conveying, mounted on a turntable, a set of hydraulically elevated booms at the topmost end of which is a cage.

**Inclinometer**

An indicator on a t.l. showing it present elevation, the maximum permissible extension at different heights for different loadings, and possibly other safety information.

**J.C.D.D.**

Joint Committee on Design and Development of Appliances and Equipment (a committee of the Central Fire Brigades Advisory Council).

**Knuckle**

The joint between the booms of a hydraulic platform.

**Leg lock**

The way in which a fireman places his leg between the rounds of a ladder to gain greater security.

**L.P.G.**

Liquified petroleum gas.

**Manifold**

A pipe or chamber with several openings.

**Monitor**

A free standing or appliance-mounted piece of equipment for delivering very large quantities of water in jet form.

**Neoprene**

A synthetic rubber.

**One/two man scale**

An indicator on a t.l. which shows the maximum permissible elevation for the current extension and the maximum extension for elevation, with one or two people on the ladder.

**Pawl**

A catch on the lower part of a ladder which prevents an extension above from running back.

**Pitch**

To erect a ladder or escape so that its head is at the point to be reached.

**Plumb**
At an angle of 90° to the horizontal.

**Plumb bob**
A device to indicate whether or not a ladder is plumb.

**Pod**
A large demountable unit used as a container for equipment.

**Power take-off**
A device to divert engine power from running the appliance to running equipment on it, such as a built-in pump.

**Projection**
The horizontal distance from a vertical line dropped from the head of a t.1. to the rim of its turntable.

**P.t.o.**
Power take-off

**Pumping appliance**
Any Fire Service vehicle carrying a pump.

**Quadrant**
A slotted, segmented guide through which an adjustable lever works.

**Rolonoff**
A type of appliance consisting of a basic frame onto which a variety of removable units can be mounted.

**Roof ladder**
A small, portable ladder used for working on roofs.

**Round**
The rung of a ladder.

**Round test**
One of the standard tests for ladders, applied to the rungs (on wooden ladders only).

**R.t.a.**
Road traffic accident

**Sheave**
A grooved pulley.

**Sheer legs**
A hoisting apparatus consisting of two or more poles joined together at or near the top.

**Shoe**
A protective, non-slip, covering for the bottom end of a ladder's strings.

**Shoot up**
To extend a t.1. with a fireman already at its head.

**Slew**
To train a t.l. (manufacturers often prefer this expression to train).

**Slip**
To remove a ladder or escape from an appliance preparatory to use.

**Special appliance**
Any Fire Service vehicle other than an ordinary pumping appliance.

**Special service**
Any operation carried out by a Fire Brigade which is not connected with the extinguishment of fire and is therefore not a statutory duty.

**Spreader**
The line or other means used to prevent the bottom parts of a step ladder from becoming too far apart.

**Strings**
The side pieces of a ladder.

S.w.l.
Safe working load

**Tachometer**
An instrument for measuring the speed of an engine or the rate of flow of a liquid.

**Tender**
A vehicle used to carry special equipment for particular situations or particular purposes e.g. an emergency tender, a foam tender, a salvage tender.

**T.l.**
Turntable ladder

**Train**
To move the head of a t.l. or h.p. by rotating the turntable.

**Trunnion**
A pin or pivot on which an object can be swung or rotated.

**Trussing**
A rigid framework used for strengthening.

**Water carrier**
A vehicle used for conveying large quantities of water to an incident where it is difficult to obtain an adequate supply otherwise.

**Water tender**
A pumping appliance carrying an increased amount of water, a built-in pump, a light portable pump, and an extension ladder or escape.

**Water tower**
An appliance such as a t.l. when used to deliver water from a height.

**Worm gear**
A gear of high reduction ratio.

# Index

# Structure and publishing history of the
## *Manual of Firemanship*

The *Manual of Firemanship* was first published in a series of nine 'Parts' (1–5, 6a, 6b, 6c and 7) between 1943 and 1962.

In July 1974, it was decided that these nine Parts should be gradually replaced by 18 'Books' and a revised format for the *Manual* was drawn up. The new Books were to up-date the information given and arrange the subjects covered in more compact and coherent groups, each group occupying one of the new Books. The following pages show the original plan, *as amended to date*. Book 5 is the eleventh of these Books to be published.

Since 1974 there have been many developments in Fire Brigade practice and equipment and in the problems which firemen may have to face. To remain an authoritative and up-to-date survey of the science of firefighting the *Manual* must take these developments into account. Not all the necessary changes can be accommodated within the format announced in 1974. The reader should therefore be aware that the structure of unpublished Books of the Manual, as set out on the following pages is subject to change. Such changes will be publicised as far in advance as possible.

The next Book planned for publication is Book 4: 'Incidents involving aircraft, shipping and trains'. This should appear in the form described.

# Manual of Firemanship

**Book 1 Elements of combustion and
extinction (published) in 1974)**

Formerly

| Part | Part | Chapter |
|------|------|---------|
| 1 Physics of combustion | 1 | 1 |
| 2 Chemistry of combustion | 1 | 1 |
| 3 Methods of extinguishing fire | 1 and | 2 |
| | 6a | 32 (III) |

---

**Book 2 Fire Brigade equipment (published
in 1974)**

Formerly

| Part | Part | Chapter |
|------|------|---------|
| 1 Hose | 1 | 4 |
| 2 Hose fittings | 1 | 5 |
| 3 Ropes and lines, knots, slings, etc. | 1 and | 7 |
| | 6a | 39 |
| 4 Small gear | 1 | 13 |

---

**Book 3 Fire extinguishing equipment
(published in 1976)**

Formerly

| Part | Part | Chapter |
|------|------|---------|
| 1 Hand and stirrup pumps | 1 | 8 |
| 2 Portable chemical extinguishers | 1 | 9 |
| 3 Foam and foam making equipment | 1 | 10 |

---

**Book 4 Incidents involving aircraft shipping
and trains (not yet published)**

Information available in

| Part | Part | Chapter | Last edition |
|------|------|---------|--------------|
| 1 Incidents involving aircraft | 6b | 4 | 1973 |
| 2 Incidents involving shipping | 7 | 1–3 | 1972 |
| 3 Incidents involving trains | 6b | 3 | 1973 |

---

**Book 5 Ladders and appliances (published in 1984)**

Formerly

| Part | Part | Chapter |
|------|------|---------|
| 1 Extension ladders, hook ladders and roof ladders | 1 | 6 |
| 2 Escapes | 2 | 3 |
| 3 Turntable ladders | 2 | 4 |
| 4 Hydraulic platforms | 2 | 5 |
| 5 Special appliances | 2 | 6 |
| 6 Pumping appliances | 2 | 1 |

### Book 6 Breathing apparatus and resuscitation (published in 1974)

| Part | | Formerly | |
|---|---|---|---|
| | | Part | Chapter |
| 1 | Breathing apparatus | 1 | 11 |
| 2 | Operational procedure | 6a | 32(V) |
| 3 | Resuscitation | 1 | 12 |

### Book 7 (first edition) Hydraulics and water supplies (published in 1975)

| Part | | Formerly | |
|---|---|---|---|
| | | Part | Chapter |
| 1 | Hydraulics | 3 | 19 |
| 2 | Hydrants and water supplies | 3 | 20 |
| 3 | Water relaying | 3 | 21 |
| Appendices | | | |

### Book 7 (second edition) (not yet published)

| | Information available in | |
|---|---|---|
| As above, plus | | |
| 4 Pumps, primers and pump operation | 2 | 1–2 |

### Book 8 Building construction and structural fire protection (published in 1975)

| Part | | Formerly | |
|---|---|---|---|
| | | Part | Chapter |
| 1 | Materials | 4 | 23 |
| 2 | Elements of structure | 4 | 23 |
| 3 | Building design | 4 | 23 |

### Book 9 Fire protection of buildings (published in 1977)

| Part | | Formerly | |
|---|---|---|---|
| | | Part | Chapter |
| 1 | Fire extinguishing systems | 4 | 24/26 |
| 2 | Fire alarm systems | 5 | 28 |
| 3 | Fire venting systems | 4 | 23 |

### Book 10 Fire Brigade communications (published in 1978)

| Part | | Formerly | |
|---|---|---|---|
| | | Part | Chapter |
| 1 | The public telephone system and its relationship to the Fire Service | 5 | 27 |
| 2 | Mobilising arrangements | 5 | 29 |
| 3 | Call-out and remote control systems | 5 | 30 |
| 4 | Radio | 5 | 31 |
| 5 | Automatic fire alarm signalling systems | 5 | 28 |

### Book 11 Practical firemanship I (published in 1981)

| Part | | Formerly | |
|---|---|---|---|
| | | Part | Chapter |
| 1 | Practical firefighting | 6a | 32 |
| 2 | Methods of entry into buildings | 6a | 35 |
| 3 | Control at a fire | 6a | 33 |

**Book 12 Practical firemanship II (published in 1983)**

| Part | Formerly | |
|---|---|---|
| | Part | Chapter |
| 1 Fire Service rescues | 6a | 36 |
| 2 Decontamination | – | – |
| 3 Ventilation at fires | 6a | 37 |
| 4 Salvage | 6a | 38 |
| 5 After the incident | 6a | 34 |

---

**Book 13**
**Contents not yet decided**

---

**Book 14 Special fires I (not yet published)**

| Part | Information available in | | |
|---|---|---|---|
| | Part | Chapter | Last edition |
| 1 Fires in animal and vegetable oils | 6c | 45(8) | 1970 |
| 2 Fires in fats and waxes | 6c | 45(3) | 1970 |
| 3 Fires in resins and gums | 6c | 45(13) | 1970 |
| 4 Fires in grain, hops, etc. | 6c | 45(6) | 1970 |
| 5 Fires in fibrous materials | 6c | 45(4) | 1970 |
| 6 Fires in sugar | 6c | 45(15) | 1970 |
| 7 Fires in paint and varnishes | 6c | 45(9) | 1970 |

---

**Book 15 Special fires II (not yet published)**

| Part | Information available in | | |
|---|---|---|---|
| | Part | Chapter | Last edition |
| 1 Fires in dusts | 6c | 45(1) | 1970 |
| 2 Fires in explosives | 6c | 45(2) | 1970 |
| 3 Fires in metals | 6c | 45(7) | 1970 |
| 4 Fires in plastics | 6c | 45(10) | 1970 |
| 5 Fires involving radioactive materials | 6c and | 45(11) | 1970 |
| | 6a | 33(VI) | 1971 |
| 6 Fires in refrigeration plant | 6c | 45(12) | 1970 |
| 7 Fires in rubber | 6c | 45(14) | 1970 |

---

**Book 16 Special fires III (not yet published)**

| Part | Information available in | | |
|---|---|---|---|
| | Part | Chapter | Last edition |
| 1 Fires in rural areas | 6b | 1 | 1973 |
| 2 Fires in electricity undertakings | 6b | 3 | 1973 |

---

**Book 17 Special fires V (not yet published)**

| Part | Information available in | | |
|---|---|---|---|
| | Part | Chapter | Last edition |
| 1 Fires in fuels | 6c | 45(5) | 1970 |
| 2 Fires in oil refineries | 6b | 5 | 1973 |
| 3 Fires in gas works | 6b | 2 | 1973 |

---

**Book 18 Dangerous substances (not yet published)**

| | Information available in | | |
|---|---|---|---|
| | Part | Chapter | Last edition |
| Alphabetical list of dangerous substances | 6c | 45(16) | 1970 |

---

Printed in the UK for HMSO
Dd 737170 C150 3/84